The Best of
Abraham Gesner

The Best of Abraham Gesner

Selected and Edited by
Allison Mitcham

LANCELOT PRESS
Hantsport, Nova Scotia

ISBN 0-88999-585-0
Published 1995

ALL RIGHTS RESERVED. No part of this book may be reproduced in any form without written permission of the publisher except brief quotations embodied in critical articles or reviews.

LANCELOT PRESS LIMITED, Hantsport, Nova Scotia.
Office and production facilities situated on Highway No. 1, 1/2 mile east of Hantsport.

MAILING ADDRESS:
P.O. Box 425, Hantsport, N.S. B0P 1P0

ACKNOWLEDGEMENT: This book has been published with the assistance of the Canada Council and the Nova Scotia Department of Education, Cultural Affairs Division.

Contents

Acknowledgements 7

Introduction 9

1. Descriptions of Some Favorite Places 17
   in Nova Scotia.
2. Descriptions of Some Favorite Places 37
   in New Brunswick.
3. Descriptions of Some Favorite Places 47
   in Prince Edward Island.
4. The Micmacs of Nova Scotia 55
5. Native People of New Brunswick 67
6. Forests of Nova Scotia and New Brunswick: 73
   With Pronouncements on Abundance and
   Varieties of Trees, Fires, Succession of
   Forest Trees and Exploitation
7. An Initial Profusion of Game and the 79
   'Wanton and Wasteful' Destruction of
   This Resource.

8. Troubles in the Fisheries of Nova Scotia   *85*
   and New Brunswick
9. Farming in Nova Scotia and New Brunswick   *99*
10. Minerals, Fossils and Oil   *105*
11. Industry in Nova Scotia   *111*
12. The Gold Fields of Nova Scotia   *117*

   Notes   *123*

Acknowledgements

I wish to thank my son, Chad, for his abiding interest in Gesner's life and work, particularly his preoccupation with fossils, for his company on many fossil-hunting trips along the Fundy coast, as well as for his insistence that I should persevere in my collection of Gesner's pieces and the writing of his biography. I must also thank my daughter, Stephanie, once more, for her support and assistance in preparing this manuscript. As well, I would like to thank the staff of the Public Archives of Nova Scotia, Mount Allison Library, Dalhousie University Library and the Public Archives of Prince Edward Island for helping me locate the material I needed. Finally, I would like to thank my publisher, Bill Pope, who immediately responded so positively to my proposal for this work and its companion volume, *Abraham Gesner, Prophet of the Wilderness*. Without the interest he expressed in publishing these two works, it is entirely possible that I might never have set them down.

Introduction

The excerpts in this book have been chosen from various works of Abraham Gesner (1797-1864), the perspicacious Nova Scotian whose reflections on so many subjects are still worth considering. Although I have chosen portions of his work which, for various reasons, most appealed to me — since *Best of* collections are necessarily subjective — I have endeavored to make my selections representative of the views Gesner expressed throughout his significant career.

Abraham Gesner was a man of extraordinary ability, perception, enthusiasm, tenacity, integrity, courage, vitality and dedication. He devoted much of his adult life to investigating and promoting his native Nova Scotia, while at the same time trying single-handedly to correct the abuses which, he clear-sightedly observed, threatened to diminish or even ruin the potential of this fabulous land. After a strenuous day of field work or doctoring, he worked late into the night setting down his observations.

In the course of his 67 years, he wrote 19 books and

reports. These dealt with geology, mining, forestry, agriculture, water power, fisheries, history, emigration, industry, medicine, scientific experiments and the native people — all subjects on which he was an authority, and often *the* authority, for his time and place.

As well as studying and reporting on most aspects of life in Nova Scotia, Gesner made extended forays into neighboring New Brunswick and Prince Edward Island, writing important treatises on both, though more detailed works on New Brunswick than on Prince Edward Island. As he repeatedly pointed out, New Brunswick had, until 1784, been a part of Nova Scotia; and, despite the political separation which took place then, nature in the two provinces bore a remarkably similar face. These parallels were particularly observable along Gesner's beloved Fundy coast where, he noted, the rock, mineral and fossil formations on one side of the Bay so frequently extended to the other.

Much of the last decade of Gesner's life was spent in the United States as a research chemist in New York. Because of his discovery of a special sort of bituminous substance (later dubbed *Albertite*) in New Brunswick's Albert County during his tenure as New Brunswick's first geologist, and the subsequent experiments he made in extracting kerosene from this "beautiful mineral,"[1] Gesner is recognized as the founder of the modern petrochemical industry.[2]

Gesner's writings are clear, fresh and frequently provocative. He said what he thought, and his views about those who held the power and the purse strings were sometimes uncomplimentary. Doubtless this was why various governmental committees often turned down his petitions, and even failed to pay him for work they had commissioned.

Gesner's works are interesting to read. They are laced

with entertaining anecdotes invariably gleaned from firsthand experience. Perhaps partly because he had travelled widely both at home and abroad, Gesner's perspective was objective. He viewed the possibilities in Nova Scotia — and Atlantic Canada in general — with one eye on Britain and the other on the neighboring United States. He reckoned that his favorite region was being shortchanged by British and Americans alike, and he protested long and loud.

Measured by the Victorian standards of his day, Gesner was an eccentric. Though a staunch monarchist, a faithful churchman (Church of England), a family physician and the father of a large family (of 11 children), he was also, at various times, horse trader, deck hand, farmer, geologist, naturalist, chemist, hunter and wilderness traveller. He was not even above helping a Yankee fishing captain, on whose boat he happened to be crossing the Bay of Fundy, to land an illicit catch, though he later decried the circumstances which made such ventures possible.[3]

Gesner loved to travel by canoe with the native people while engaged in his various surveys, and even employed several Micmacs to stuff animals in the attic of his St. John house for his own museum of natural history. Of this venture, his son George remarked:

> The moose, deer, caribou and larger animals were all stuffed and preserved at St. John, at his residence on McNab's Hill, and in setting them up several Micmac Indians were employed, and almost any evening during the progress of the work a group of them could be seen sitting about the fire, at the end of a large attic, smoking killikinick and chatting in their soft tongue before wrapping themselves in their blankets and going to sleep.
>
> These Indians, many of whom were Dr. Gesner's

guides in his surveys, and who were good judges of the natural attitudes and appearance of the stuffed animals, were very capable assistants in setting them up.[4]

Gesner risked his life time after time — early and late — on ventures which rarely paid off in cold cash, but which he did not regret. He played the flute and violin at home and, apparently, according to his associate, Moses Perley, the bugle,[5] on wilderness lakes. He seemed equally at home with sophisticated intellectuals, farmers, seagoing men, the audiences he captivated and the native people. Gesner was, in fact, so popular with the native people that they called him "wise man,"[6] and, somewhat to the exasperation of Moses Perley, who had engaged Malecite guides for a wilderness expedition in New Brunswick, directed him to the best fishing spots.[7]

Gesner was warm-hearted, even-tempered and loved a good story, but he did not drink "spiritous liquors," remaining abstinent in an age when rum in particular was routinely issued to Maritime sailors and lumberman, and when even the Halifax poorhouse was provisioned with beer and wine.[8]

Despite the fact that he lived most of his life in the Maritime Provinces which were then chiefly wilderness, Gesner was in close contact with some of the best minds of his age. His observations on geology and his public demonstrations of his experiments in the initial stages of the petrochemical industry were attended to and admired abroad in the most advanced intellectual circles — though sometimes ignored or treated disparagingly closer to home.

In London, his geological studies were honored, his maps and articles published in prestigious journals, and he was made a fellow of the Geological Society. The foremost

British geologist of the time, Charles Lyell (later Sir Charles) on his 1841-1842 travels in North America, sought out Gesner as the authority on the geology of the Fundy coast, and subsequently cited him in his book on his 1842 summer's observations.

In New York too, in the 1850s and early 1860s, Gesner was accorded similar recognition. His work was praised at length in issue after issue of *The Scientific American*, and the book which brought him most fame universally — *A Practical Treatise on Coal, Petroleum and other Distilled Oils* (1861) — was published by Baillière Brothers, 440 Broadway. This prestigious publishing house, which had branches abroad, brought out this ground-breaking work simultaneously in New York, London, Paris, Melbourne and Madrid. It remained for many years the most authoritative treatise on the subject.

However, Gesner's observations on the Atlantic Provinces, on Nova Scotia in particular, have never been given the attention they deserve. These commentaries, vivid and firsthand descriptions of the environment of this region published between 1836 and 1862, show a significant Maritime writer at his best. They are likely to hold greater interest for most contemporary readers than the scientific treatises which brought him brief renown late in life. It is excerpts from these regional Maritime works, then, which dominate this volume. Their sources are identified in endnotes.

After reading these pieces, some may find that Gesner rivals his well-known contemporary from Windsor, Judge Thomas Chandler Haliburton, with the clarity and charm of his prose. It is interesting to note that, although Gesner agrees with Haliburton's complaints about the general lack of energy and expertise of Nova Scotians, in comparison with the energetic and successful enterprise of their Yankee neighbors, his criticisms are couched in gentler terms.

Gesner tends to make excuses for the shortcomings of his compatriots.

The story of the ups and downs of Gesner's life is fascinating, but it is told elsewhere — in the companion volume to this work — *Abraham Gesner, Prophet of the Wilderness*.[9] *The Best of Abraham Gesner* aims simply to bring to light a cross section of the most interesting and readable parts of Gesner's prolific literary output. His publications have long been out of print. Libraries which house the extant copies do not allow them out on loan: replacements are too expensive and difficult to find. It is time for Gesner's work to be brought before the public once more.

1
Descriptions of Some Favorite Places in Nova Scotia

The Bay of Fundy

The Bay of Fundy was visited by DeMonts in 1604 and called by him La Baye Françoise. It is 100 miles long, and 40 miles wide, and nearly separates Nova Scotia from New Brunswick. It terminates in two smaller bays, namely Chignecto or Cumberland Basin, and Minas Basin. The latter is upwards of 50 miles in length, its greatest breadth being about 30 miles. It is a beautiful sheet of water, expanded into the central part of the province; and, with its numerous rivers, it affords great facilities for trade with the United States and New Brunswick; but not with Halifax, with which, as already proposed, it should be united by railway. The extensive shores of this basin afford every advantage for ship-building; and at many points the land is uncleared, and supplies of ship-timber may be procured from the bordering hills.[1]

Seen from any of the neighbouring hills, the landscape is one of the most splendid and varied in America. The shores of the land-locked Minas are indented by river mouths and caves, along whose banks cultivation has spread out its mantle of green. In autumn the rescued marshes, which had yielded their crops of clover and wheat, are covered with droves of cattle: while, at high water, vessels appear to be sailing among them in the display of some great exhibition. At a thousand points the native forest still reaches the cliffs, or the beveled edges of land; and, far away, the wilderness is dotted with the little clearings of the 'new beginner.' The rivers of King's and Hants counties appear like a large hand, with the fingers laid upon the fertile sides of the valleys, opening towards the shore. Villages, and long lines of comfortable white farm-houses, are stretched along the courses of the rivers; and here and there a church or chapel sends up its sharpened spire ...

Chignecto Bay is also a fine sheet of water, terminating in Cumberland Basin, which separates the county from New Brunswick. Cumberland Basin (Beau Basin of the French) is a safe harbour; but from the great elevation of the tides, vessels lie aground at low water ...[2]

These bays are remarkable for the elevation and rapidity of their tides, which, at the extremities of the estuaries, rise 75 feet, and in the narrow straits the currents run at the rate of 10 miles per hour. This great periodical elevation of the surface makes docks of all the creeks and rivers; and vessels are thoroughly repaired to their keels, between the high and low tides. They may also be placed in situations where they will be left dry 16 hours out of 24.

The advantages arising from this extraordinary influx and reflex of the sea are by no means inconsiderable. The extensive and fertile marshes are the gifts of the tides, which fill the estuaries with the fine sediment abraided from the rocks of the coast. Wares and nets are stretched along the beaches; and, at low water, the fish taken in them are removed in carts.[3]

Cape Split, Lithograph, W.H. Bartlett, J. Consen, from N.B. Willis, **Canadian Scenery** (Volume II), London: Virtue and Company, 1842.

Cape Bloomdon and Parrsboro, W.H. Bartlett, R. Brandard from **Canadian Scenery** (Volume II).

At the head of Cobequid Bay the flood tide is preceded by an immense tidal wave, or bore, which, at spring tides, is sometimes six feet high. At low water nearly 60 square miles of sand, shingle, and mud flats, are laid bare: the flood rises more rapidly than the water can advance, and the result is the formation of a splendid wave, sometimes more than four miles long, which rolls over the flats and quicksands in a sheet of foam, and with the roar of thunder, washing away or burying up everything before it. Vessels lying with their broadsides to the bore are rolled over — their masts are broken, and they are left half buried in the shingle: the skill of the pilot is, however, equal to this danger, and accidents occur but seldom. The Petitcodiac, and other rivers of Chignecto Bay have their bores; but they are less powerful than the mighty flood of the Shubenacadie.

The scenery at the entrance of Minas Basin is bold and picturesque. Blomidon, Cape Split, Partridge Island, and Cape D'Or, with their lofty facades of trappean columns and overhanging cliffs, at once arrest the attention of the traveller, as he glides between these mountain masses, urged forward by the incredible fury of the tide. The submarine besalts of Cape Split and 'Dory,' even in a calm day, break the surface of the water into frightful eddies and sheets of foam. The former is the maelstrom of the Bay; still the navigation is safe when intrusted to experienced pilots, who know their positions in the greatest darkness and thickest fogs by the peculiar sounds or 'routes upon the shore,' and 'racket of the rips.' The passage of a field of ice over one of these rips in the winter season is at once a grand and amusing spectacle. It was called by an American captain 'about the greatest muss of a hoddy-doddy in all creation…'[4]

Partridge Island

Four miles eastward of Cape Sharp, Partridge Island, so justly celebrated for its numerous minerals and picturesque scenery, rises in lofty grandeur from the Bay. Its name has probably arisen from the numerous broods of partridges reported to have frequented its surface. These however have long since disappeared, leaving the fox sole occupant of its elevated area. This island is northward of Blomidon about six miles, and separated from it by the narrow outlet of the Basin of Minas called "The Gut." These strong collections of basalt, may be compared to the pillars of a gateway, and being more unyielding in their natures, have resisted the impetuous tides more faithfully than the shale and sandstone which they defend. While Blomidon protects the softer rocks of Cornwallis from the encroachments of the aqueous element, Partridge Island and Cape Sharp, on the opposite side of the Gulf, stand like giants guarding the coast of Parrsboro. The great body of water rushing forward to fill the Basin of Minas during the flood tide is thus compelled to pass this narrow opening, and rushes with fearful rapidity along its confines. Through this contracted but deep channel, the tide runs eight miles an hour during the spring tides; and when it is considered that it rises upwards of 50 feet high in the short space of six hours, its violence cannot be surprising. Often in calm weather, when vessels have been prevented from anchoring by their distance from the shore, they are seen passing along with great speed stern foremost, through the "Gut," over the rippling surface of the Basin. Frequently a single flood will sweep them 50 miles, either towards or from the place of their destination. It is on this account that the harbour of Partridge Island is so extremely advantageous; the peculiar direction of the currents, with a light air of wind, almost always allows them to make that port, where they are often collected in great numbers. Partridge Island not only offers a retreat from the fury of the tides, but also in tempestuous weather affords a shelter both on its east and west sides. A fine navigable river at hand, where a thousand vessels may lie in safety, opens in the rear; here, during the cold and blustering months of autumn, their masts are seen rising almost among the clusters of spruce, like leafless pines among the underwood...[5]

On the west side of Partridge Island, the basaltic trap deposited on amygdaloid, forms a sublime and stupendous cliff, 300 feet high. At some places it is undermined, and hangs frowning over the visitor who ventures to stroll among the fragments at its base. Beneath the most lofty part of the precipice, a cone-shaped insulated mass of amygdaloid rises to considerable height only a few yards from the base of the more exalted summit of the island. Even this limited portion of the rock is beautifully topped with regular basalts. At the southeast extremity of the island, several small detached and grotesque rocks form a kind of rampart in front of a rugged point.

These rocks are beautifully crowned with low evergreens, above which the spear-shaped top of the spruce mocks the fury of the waves beneath. Viewed from the summit of the lofty precipice, or from the beach below, the whole group affords a grand and beautiful landscape. On the east side of the island, a large mass of basalt has made an attempt to fall; but being arrested in its descent has left on the earth above a deep and narrow chasm into which access may be had by carefully descending a short distance down the border of the 'escarpe' above. This chasm has probably been produced by an earthquake, and there are numerous evidences in the Province of those dreadful events...[6]

At Partridge Island the upturned strata are a little

retired from the shore. Upon their most prominent ridge stands the blockhouse, while between it and the basin a narrow space is sufficiently level to afford an easy communication with the harbour, and to accommodate the inhabitants of the beautiful village standing upon its border. Near this romantic spot there is an extraordinary locality. The remains and impressions of large flags, and other aqueous plants, are found in strata of blue shale. Several other small vegetables have been observed imbedded in this rock; and their appearance indicates that they were the growth of low and moist grounds. Immediately alongside the strata containing these antediluvian records, and in immediate contact with their branches and leaves, the rock is filled with myriads of fossil shells which have been obtained in great perfection and beauty.

These shells belong to a species of 'mytilus edulis,' or common fresh water muscle; none of their kind ever inhabited the saline waters of the ocean, and their home was at the bottom of some lake, pool, or rivulet, whose waters flowed from higher grounds. Here the productions of the soil, and the testaceous inhabitants of the water, have been buried side by side. Here the animal and vegetable creations have been laid up together in the vast cabinet of Nature; here is testimony of the most extraordinary kind, proving the changes which have taken place upon the earth, since it rose from the dark and mighty deep. And great must the overthrow have been when the inhabitants of the peaceful lake or brook, and the plants which blossomed upon its margin, were doomed to be imprisoned in the solid rock, where their remains still appear, forming the monuments of their former vitality and the awful shock that hurled them upwards above the level of the waters. Had the phenomena which accompanied the events recorded at this place been inscribed by the hand of man, many would ridicule them as idle phantoms: but as the most faithful Historian has placed the objects of the narrative before our eyes, we should begin to know how limited our knowledge is of what is past.

A bed of coarse breccia forms the east side of the mouth of Partridge Island River; the grey sandstone and black shale, in perpendicular strata, compose the shore as far eastward as Clarke's Head, where it is succeeded by a bed of diluvium; and afterwards by the new red sandstone, which is finally overlaid by a mass of trap ... At this locality all the rocks have been under the influence of heat during the formation of the greenstone in the neighbourhood; and there are few places in the Province where so great a variety of minerals has been produced.[7]

Opal and semi-opal are also among the interesting specimens of Partridge Island. We have obtained two small nodules of the former — both resemble pieces of wax. Both varieties are greatly improved by the labour of the lapidary, and, when they are cut and polished, afford beautiful gems. This mineral was much esteemed by the ancients. Nonias, a senator, suffered banishment rather than relinquish to Mark Anthony a precious opal.

Of all the minerals for which Partridge Island has been celebrated, none is more admired and carefully sought for than the amethyst. This gem is now becoming scarce, and nothing but the downfall of the cliff will bring it to view. Monsieur DeMonts, a leading character among the French emigrants to this country during the reign of Henry IV, was so much pleased with these brilliant crystals that he conveyed a number of them to France where they were received by the King and Queen as a token of his loyalty. Many similar specimens have been transported to Europe, where they are much admired.

The north side of the island descends with a somewhat

Parrsboro from the Water, B.F. Nutling, Miss Jeffrey, from Gesner's Remarks on **The Geology and Mineralogy of Nova Scotia**, Halifax: Gossip and Coade, 1836.

gradual slope. About half the distance from its summit towards the base, a large vein of the magnetic oxide of iron breaks through the soil; the ore is of a good quality, and sufficiently abundant to supply a smelting furnace. The waters of a spring in the neighbourhood are impregnated with the ferruginous oxide and possess medicinal virtues superior to far more celebrated pools.[8]

The advantages to be received by residing in a part of the country where minerals are abundant, were not the least of those motives by which we are led to occupy our present domicile; and although it is very uncertain how near the period may be when an unwilling migration must take place, we cannot hesitate to affirm, that over and above the Mineralogical inducements, Partridge Island offers a neat and airy village, on the very border of the

Basin, attended with every facility for sea bathing, and the pleasure to be derived from a kind and hospitable society; it offers every inducement to the relaxed invalid, and all who can enjoy a pleasant summer retreat. Here the Mineralogist will find the objects of his study beside him; the calculating merchant would be invigorated from new labours; and the fair daughters of our land would receive another charm from the refreshing influence of a pure and wholesome atmosphere.[9]

Cape Breton

The Island of Cape Breton ... forms a valuable part of the Province and equals the main land [sic] in natural resources. Abounding in coal, limestone, gypsum, and other minerals, the soil is fertile, except at its northern and mountainous extremity, and along a part of its Atlantic side between Isle Madame and Gabarus Bay. There is also some boggy ground in the interior. The southern and western shores abound in harbours, and the deep Bras d'Or penetrates the very centre of the Island, extending its navigation into the remote forests and to inexhaustible strata of mineral fuel. The waters also teem with fish of every variety, and the inland districts still afford supplies of excellent timber.

Cape Breton, originally called Isle Royale, was first discovered by John Cabot, a British navigator. The early 'voyageurs' to it were from Bretagne, in France: hence the origin of its present name. France always considered this Island as the key to the St. Lawrence, and she expended 30,000,000 of livres in the fortification of the capital, Louisburg, so called in honor of her sovereign. During its occupation by the French, it exported 5,800,000 quintals of fish annually, and 600 vessels were employed in its trade and fisheries.

I had an opportunity of visiting the ruins of Louisburg,

the ancient capital of Acadia, in November last ... The light-house on the northern side of the entrance of the harbour stands on a bold rocky cliff, once occupied by a strong battery. The dilapidated walls of the great battery of 40 guns on the northern side of the harbor, and another on the opposite shore, now appear like natural mounds, being covered with clover and other grasses. The little island at the harbor's mouth has yielded to the operations of the waves, and a part of the fortification has fallen into the sea, but the walls, entrenchments, and the town of Louisburg — the missiles — the blown up batteries — levelled city, and the bleached bones of the dead, now seen mingled with the soil, best show the sacrifices made to secure the advantages of the situation. The arched places of arms, and bomb-proofs of the citadel, are still entire. Three of them are sheepfolds — another is occupied by a fisherman for a cabbage cellar, and time has incrusted the ceilings of all with small stalactites. The foundations of the barracks, chapels, the nunnery, hospital, and other public buildings, are still perfect; and the cells of the prison are almost unbroken, as is also the kiln of a large brewery. The present inhabitants are supplied with water from the Governor's well, and the walls of some of the buildings and chimneys are 12 feet high ...

Excepting the west coast, the Island abounds in fine harbours, and the shores are peculiarly favorable for the occupation of the fisheries. Since the fall of Louisburg, Sydney has been the capital. The entrance to the harbour is wide, and the shores present chains of small farms, cultivated by Highlanders. The rocks all belong to the coal series, containing coal, and a great variety of the beautiful fossils belonging to the carboniferous group ...

Arichat has long been celebrated for its exports of the produce of the sea. The present population will exceed 3000, chiefly Acadians. The trade is carried on principally by merchants from Jersey who employ the inhabitants in taking and curing the fish. In 1828, 39,200 quintals of dry cod, and 12,559 barrels of pickled fish were exported ...

The Scotch settlement continues to the excellent harbour of Port Hood, the county town for the northern district. Still farther on the shore becomes bold and precipitous ... The banks of Marguerie, or Salmon river, 50 miles north of Port Hood is occupied by flourishing settlements of Acadians, who are also scattered along the coast as far as Cheticamp, where the Jersey merchants have another fishing station. The remaining part of the coast is but thinly inhabited, and there is no shelter for vessels.

A large tract of country in the northern part of the Island is occupied by lofty mountains of granite and trappean rocks. This tract has never been explored, and I have only had an opportunity of making a few hasty observations along a part of the coast. The whole shore from Cape St. Lawrence to Cape North, and thence to Inganishe [sic], presents perpendicular cliffs of granite and other igneous masses, which descend into the sea without a beach border; and at numerous sites a landing cannot be effected, even in calm weather. Against the shelving cliffs of granite and basaltic columns of trap, the sea dashes with terrific violence. This part of the Island is the highest land in Nova Scotia. Some of the mountains exceed 1200 feet in height. Between them there are deep gorges, flanked by almost vertical precipices reaching from the level of the sea to the summits of the mountains, some of which are levelled off at their tops. The ice and snow of winter form glaciers, the debris of which is seen in the valleys. The scenery is majestic beyond any in the Province ...

The position of Cape Breton in regard to the fisheries of the North American coast is peculiarly favorable. To this may be added the number and safety of its harbours — the abundance of fish that frequent the coasts, and its advantages for capturing seals. The coal mines also are sources of wealth, and the interior lands abound in timber. With all these advantages, the Island will bear no comparison with other, but less favored parts of North America. Its resources are not known nor appreciated abroad, and such as have been discovered are very inadequately improved. The coasts are much exposed to smugglers and trespassers upon the fisheries; the prosperity of the Island, therefore, depends as much upon its maritime protection as upon the encouragement offered to internal industry.[10]

Canseau [Canso]
Canseau, situated at the south-easternmost part of the province, is an excellent harbour, open at all seasons of the year, except in the spring, when, like Chedabucto Bay, it is often filled with drift ice from the Gulf of St. Lawrence: yet it is seldom unnavigable more than two or three days at a time. It is formed by St. George's, Durell's and other small islands, between which and the main land [sic] there is a deep channel, with good anchorage. This harbour, although not very capacious, is the resort of vessels during gales from the westward: as a fishing station it is unrivalled.[11]

The remains of forts are still to be seen on Green Island, and the adjacent shores; and the stone axes, spear and arrow heads of the natives are found in the soil and on the sides of the harbour.

St. Andrew's, Whitehead, Raspberry and Whale Islands, and Dover Harbor, all afford shelter. The coast in this quarter is formed of white granite, which has been sculptured by the waves into the most fantastic forms; and, from being undermined, large shelving masses slope towards the sea like the roofs of houses.[12]

Canseau was the first part occupied by the English,

French and Spanish fishermen; and from being near the key of the Gulf, it became a place of severe contest between the claimants of North America. The native Indians also held this station in high estimation on account of its advantageous position and fisheries.[13]

The strait, or Gut of Canseau as it is commonly called, is the grand thoroughfare of all the Provincial and American trade to the Gulf of St. Lawrence. This remarkable channel is about a mile wide, upwards of 20 fathoms deep, and 15 miles in length. The current usually runs at the rate of four and a half miles an hour, and sometimes several days in one direction, according to the winds. The scenery is magnificent, and, during the summer season, it is enlivened by fleets of vessels of almost every description. This strait separates Cape-Breton from Nova-Scotia, to which it was no doubt attached at some distant former period: the geological evidences are such as indicate the breaking through of a narrow isthmus by the operation of powerful currents. It now opens into spacious harbours, abounding in fish of various kinds.

The narrowness of the passage renders it capable of being defended; and the progress of fleets may be arrested by batteries upon the shore. The number of American fishing vessels that pass through the strait, going to, and returning from, the Gulf, has been computed at 3,000 per annum: adding to these the number of coasters, fishermen, and timber-ships of the provinces, with those employed in the Pictou coal trade, the aggregate is immense.[14]

Halifax

The harbour of Halifax (in Lat. 44 40'N. Long. 63 60'W.) is universally admitted to be the best in America. No sooner was it discovered than the British and French fleets made it their point of rendez-vous, in the early struggles of the two powers to occupy and hold the country. It was also a favorite resort of the Micmac Indians, who, for many years, continued their barbarous warfare against the first colonists who settled upon its shores. The lakes of Dartmouth, and the Shubenacadie, afforded them an easy communication with the Bay of Fundy. The fisheries and hunting grounds in this quarter were also highly prized.[15]

The fish market of Halifax is unrivalled, and affords its dainties in the severest cold of winter, besides sending supplies to New York and Boston by the Atlantic steamers, and the American packets.[16]

The navies of all Europe might enter this harbour, which is accessible at all seasons of the year. It opens into the Atlantic from the north, and, after extending fifteen miles, terminates in a beautiful land-locked basin, where whole fleets may ride at good anchorage. Its wide entrance between Sambro Light and Devil's Island, is almost free from danger. MacNab's Island, on the eastern border, is three miles long: between it and the main land [sic] is the Eastern passage, frequented by the smaller vessels. The entrance to this port is well lighted, and buoys are fixed upon all the shoals. A fine deep channel stretches towards Margaret's Bay, called the North-west Arm, which renders the site of the City of Halifax a peninsula.

From the numerous advantages of this port, it has been wisely selected for the capital of the province, which stands on its western side, surrounded by fortifications, and defended by George's Island, a bold elevation in the middle of the harbour. Halifax is the first safe port that can be reached on the continent of North America at all seasons of the year, after leaving Great Britain. It has therefore been selected for the first landing for the steamers from Europe, and must be the terminus of the great railway of the provinces, which, with a railway to Windsor, would render Halifax one of the greatest commercial cities in America.[17]

Mahone Bay

Mahone Bay, justly celebrated for its beautiful scenery, is upwards of 12 miles in diameter. The Tancook Islands, at its entrance, break off the sea to the south-east. The Harbour of Chester is also protected. The whole shore abounds in river mouths and inlets of deep water. The islands thicken as the maritime traveller advances, and assume infinite variety in shape and feature. A few are thinly inhabited — others are covered with evergreens. The whole form a labyrinth, whose attractions can scarcely be imagined or described. The principal streams are Gold and Mushamush rivers. The former springs from a small lake, and, on account of its fine salmon, has long been celebrated by the disciples of Isaac Walton.[18]

Lunenburg

The harbour of Lunenburg is separated from Mahone Bay by a narrow peninsula, and is accessible for ships of the

Lunenburg, *The Illustrated London News*, October 5, 1861. Courtesy of Special Collections, Killam Memorial Library, Dalhousie University.

largest class. Its borders were settled by Germans and Swiss in 1751. The fertility of the soil, the fisheries, facilities for commerce, having been improved by a sober and industrious population, have made the wilderness in this quarter 'to blossom as the rose.' The town of Lunenburg stands upon the north-east bank of the bay. At its commencement it was much exposed to the attacks of the Indians, and remnants of the block-houses and palisades still remain, that were erected by the first immigrants for its defence.[19]

LaHave

LaHave, six miles westward of Lunenburg, was taken possession of by the French in 1613. In 1634 La Tour obtained an extensive grant of land along its banks; and his fort, at the entrance of the bay, still to be seen, was the theatre of many tragic events. At the entrance of this admirable harbour there are a number of islands, affording the necessary shelter against the winds and waves of the boisterous Atlantic. Some of these are mere islets, or naked rocks. Some are tufted with verdant spots, while the larger and more fertile of the group, are shaded by forest trees and shrubbery. The noble and most romantic river enters the outer harbour, through a narrow passage and is navigable to the distance of 15 miles: a bar at its mouth has 21 feet water at full sea. Three miles above, at the ferry, the river is three quarters of a mile wide, and from three to six fathoms deep. Eighteen miles from its débouchement, it passes over a fall of 20 feet. Six miles farther up it has another cataract of surpassing beauty. This fine stream takes its rise at the Kempt lake; and its extreme sources touch the head of the Gaspereau, emptying in the opposite direction into Minas Basin. The main trunk of the river, and its lakes, were the route of the Indians in former times, when they transported their light bark canoes up the rapids and across the portages to the Gaspereau lakes at Horton, and thence to the Bay of Fundy. The river is now occupied by numerous saw mills, and therefore its salmon fishery has been nearly destroyed. The whole scenery is of the most fascinating character. The uncultivated intervales are covered with shrubbery; and, on the hills, the tall ... pine, peering far above all its competitors, waves gracefully in the air. The lakes also have their little islets, shaded by the maple, the beech, or the oak. Nature, in all its wild luxuriance, still holds uncontrolled dominion; and the bear and moose wander through the forests fearless of the hunter's skill.[20]

The South-West Coast
The south-west coast of the province is low, and abounds in caves and creeks, skirted by low marshes. It is defended by islands of every form: in Argyle bay alone their number is said to be upwards of 300. These islands, with the lakes, creeks, and marshes, of the main land [sic], present an infinity of form and beauty. Anchorage is offered for vessels of all sizes; and, besides excellent fisheries, every convenience is offered for ship-building.

Tusket River, opening into the above bay is navigable eight miles. Above the tide it connects a chain of lakes, which may be traversed in boats to the distance of 30 miles. The whole interior of this part of the province is interspersed with lakes from one to ten miles in length. The peculiar features and beauty of the Tusket made it the resort of the Indians, who, with their light barks, crossed thence to lake Rossignol and to the Sissiboo, emptying into St. Mary's bay. There is a tradition that the aborigines formerly assembled here to offer sacrifice to the Great Spirit, and the traces of their hieroglyphics still remaining upon certain rocks are such as corroborate that opinion: but it was to these interior retreats and fastnesses that the

Indians and frequently the French neutrals made their escape, when they were defeated upon the seaboard, thus the sites of old encampments, known by the presence of apple trees and Indian relics still found in remote situations.[21]

Brier Island

This island forms the most westerly extremity of the Trap Formation, and is separated from Long Island by a narrow channel, through which the tides pass with great rapidity. From that cause, and the exposed situation of its rocks to the open sea, it suffers much from the destructive powers of the elements, and more than the Islands in the Basin of Minas, or Mahone Bay, which are somewhat sheltered from the violence of the waves. On the south side of the island, and near the entrance of the channel from St. Mary's Bay, the rocks have been worn away, and beautiful cliffs of regular columnar basalt are exposed to the ocean. The columns form long ranges of pillars, like the steps of stairs, reaching from the sea below to the precipice above, against which the waves often dash with fury, breaking down the notched ridges and pedestals forming its base. These pillars are in general hexagonal, although some are enclosed by seven sides. Their articulations occur at short intervals. This circumstance renders the rock more liable to be broken down than it would be were the columns of greater length. On this side of the Island the basalts extend outwards beneath the sea, forming a submarine causeway, called *the Bar*, over which the tide and waves rush with great force, forming and breaking over the impediment thus placed in their way; the sea sends forth hollow sounds like those of distant thunder, and in calm weather may be heard several miles off. On the western side of the island, and near the lighthouse, the rocks attain a greater elevation, although their columnar arrangement is not so manifest. At low water the red sandstone was seen cropping out beneath the trap, thus confirming an opinion already advanced, and supporting a fact of considerable importance. In comparing specimens of basalt from Brier Island with those from the Island of Staffa, they were found very similar, and no important feature was wanting, to identify the rock of Nova Scotia with those of the celebrated Fingal's Cave ...

From the exposed situation of Brier Island to northern gales and thick fogs, the soil is unproductive; but what nature has withheld in vegetation she has supplied in fish, which are excellent in kind and quality ... We cannot forget an opportunity afforded for surveying this island in 1821, although the circumstances connected with our visit at that time were not of the most pleasing kind. On the last of December of that year, on our way to the West Indies, the vessel in which our lot was cast, was overtaken by a violent gale of wind: she soon became a perfect wreck — the crew frozen and exhausted. Fortunately a change of wind drove the crazy bark into the Grand Passage. There had been a cargo of twenty horses upon the deck, but when we landed only five remained, and they had been dragged ashore in the turns of the cable, which had washed overboard, and so encircled them as to prevent their escape. Nor should the kindness of Charles Jones, Esquire, and his family be forgotten; to them we feel greatly indebted, and the marks of frost still remaining upon our lower extremities, will not allow the circumstance to flee our memory.[22]

2
Descriptions of Some Favorite Places in New Brunswick

General Observations

There is great diversity in the appearance of the province in regard to its surface. Along the coast of the Bay of Fundy, and extending northward to a distance of 30 miles, there is a tract of hilly country, occupied by deep and narrow ravines, which give the surface a mountainous appearance; but few of the hills attain any considerable degree of elevation, nor are they such as would materially retard the progress of cultivation. Watered by numerous rivulets descending from the higher grounds, the ravines and valleys vent the smaller streams, which being collected in rivers are frequently poured into the bay over beautiful cataracts or boisterous rapids. In this district there are many large tracts of naked rock, and numerous peat-bogs, or mossy swamps, which could only be reclaimed by a dense population, and in an advanced state of agriculture. Although there are many fine belts of intervale along the streams, and some patches of good soil on the hills, this division of the country, like the south side of Nova Scotia, is not well adapted for agriculture. The scenery is wild and picturesque, the bold cliffs or ragged precipices, the deep valleys, the quiet lakes and the dashing waterfall, are sometimes presented at a single view. The close forests of hill and valley appear in summer like green waves, rising in succession above each other. Dotted on their sides by the log-house and clearing of the settler, they declare at once the still infant state of the colony and the slow progress of husbandry.

The whole north-eastern side of New Brunswick, from Bay Verte to Bathurst, presents a low and level surface, almost unbroken by hills. The country at many places is uneven; but there are few steep acclivities, except those that have been produced by the action of water upon the beds of the rivers and other streams. Extensive marshes, intervales, and floating peat-bogs are somewhat peculiar to this part of New Brunswick. The above tract extends in a south-west direction to the River St. John. It is the region of the great coal-field of New Brunswick, and occupies an area of 5,000 square miles. Although there are numerous parcels of land too light and sandy to be very productive, the soil in general is good, and many tracts are of a superior quality.

There is another tract of country, extending from the Meductic Falls on the River St. John to the Acadian settlement at Madawasca, and thence in the north-east direction to the Bay Chaleurs [sic] and Restigouche. This district is mountainous, and embraces a part of the chain of highlands to which the Treaty of 1783 referred in reference to the boundary between the Province and the State of Maine.

Viewed altogether, the face of the country is greatly diversified, and exhibits almost every variety of scenery. It is indeed difficult to form a correct idea of what the appearance of the wilderness region will be after its surface has been partially cleared of its burden of timber and its level alluviums changed into fertile meadows. At many places in the wild woods there are noble streams passing through the intervales and winding along their courses through lofty groves of ash and elm. Standing along the borders of these rich fields of wild grass there are sometimes abrupt rocky cliffs crowned with spruce and other evergreens; but so close is the forest that it is only from the summit of some naked eminence that the natural beauties of the country can be perceived or its future appearance be anticipated.[1]

The highest mountains in the Province are situated at the source of the Tobique, Upsalquitch, and Nepisiguit rivers. Blue Mountain, Ox Mountain, Pot Mountain, and Bald Mountain, of this range, will exceed 2,000 feet in height. This highland district affords some of the most sublime scenery in the province. The summits of the mountains are most frequently naked. In some of the deep chasms and ravines, at their northern bases, where the rays of the sun are obstructed, the snow does not disappear during the summer, and in the spring glaciers sometimes descend, sweeping the woods before them downwards into the valleys below.

The streams pass through narrow and tortuous channels, frequently overhung by stupendous cliffs; and the water dashing from fall to fall, is finally lost in wreaths of spray and foam in the more quiet streams of the lower ground. From the mountain tops nothing is to be seen in the foreground but vast masses of shelving rock, which frequently overhang the tops of large trees that have fastened themselves to the declivities, or stand erect from the bottoms of the gorges ...

There is here a tract of country at least 300 miles in circumference upon which there is not a human dwelling;

and the presence of the industrious beaver is evidence that the Indians seldom penetrate so far into the wilderness.[2]

Rivers
There is perhaps no country in the world of the same extent that enjoys greater facilities of navigation than New Brunswick. All its large rivers are navigable for ships, and its smaller streams afford safe passage to boats and canoes.

The St. John is the largest river to the province. It was discovered by DeMonts on the 24th June, 1604. By the native Etchemins it was called the Looshtook, or Lahstok (Long River) ... It received its present name from having been discovered on St. John's Day ...

By winding its way along the segment of a large circle, it traverses the country to a distance of 500 miles, until it finally empties itself into the Bay of Fundy ..."[3]

Between Fredericton and the mouth of the St. John, the main river resembles a lake. The tide flows to Chapel Bar, four miles above the capital, and seldom rises over 15 inches. The noble stream is now spread out into small bays, and inlets communicating with lakes, along its margin. In descending, the valley is greatly enlarged, and its whole area is occupied by extensive tracts of alluvial soil, islands, ponds, and creeks, through which the majestic St. John sullenly winds its way, bearing upon its bosom the steamboats and numerous craft of the river. The alluvial banks, as well as the higher grounds, are extensively cultivated. The rich meadows are decorated with stately elms and forest trees, or sheltered by low coppices of cranberry, alder, and other native bushes. Through the numerous openings in the shrubbery, the visitor, in traversing the river, sees the white fronts of the cottages, and other buildings; and, from the constant change of position, in sailing, an almost endless variety of scenery is presented to the traveller's eye. During the summer

Saint John from the Signal, W.H. Bartlett, C. Consen, *Canadian Scenery* (Volume II), 1842.

season, the surface of the water affords an interesting spectacle. Vast rafts of timber and logs are slowly moved downwards by the current. On them is sometimes seen a shanty of the lumberman, with his family, a cow, and occasionally a haystack, all destined for the city below. Numerous canoes and boats are in motion; while the paddles of the steamboat break the polished surface of the stream, and send it rippling on the shore. In the midst of this landscape stands Fredericton, situated on an obtuse level point formed by the bending of the river, and in the midst of natural and cultivated scenery ...[4]

The river [12 miles above Woodstock], with its wooded islands and high terraced border, surmounted by cultivated uplands, is well calculated to strike the eye of the traveller after he has ascended from the tamer scenery

Falls on the Saint John River, W.H. Bartlett, C. Consen, **Canadian Scenery** (Volume II), 1842.

Split Rock, Saint John River, W.H. Bartlett, C. Consen, **Canadian Scenery** (Volume II), 1842.

below. The St. John is here a furlong wide, and the stream runs smoothly along at the rate of six miles an hour. The timber on the uncleared lands consists of spruce, fir, cedar, and pine, intermixed with birch and maple. The islands are covered by the different varieties of hard wood [sic] and butter-nut.[5]

There is not a river in America of the same extent that has so narrow an outlet as the St. John. From Grand Bay to the falls, a distance of four miles, this noble stream passes through a crooked channel, at many places not exceeding 250 feet in width, while in the interior of the country the stream will average from one to three miles in breadth. The rocky shores of its outlet have not been worn down and scooped out, as is common in all rivers giving passage to great quantities of ice. On the contrary, they appear to have been separated from each other at a period comparatively recent; and the gorge through which the river now passes at Indian Town appears like a deep fissure opened by some sudden movement in the earth. It is probable that the St. John had formerly two mouths, one opening from the Kenebecasis down the present site of the marsh, and the other opening from Grand Bay through the Manawagonish...[6]

The harbour of St. John is neither very spacious nor commodious. From its shallowness and the violence of the current, large ships cannot enter it at low-water. Those disadvantages are in some degree compensated by the elevation of the tides, which are very favourable to shipbuilding and the transportation of timber. The *débouchement* of the river is between perpendicular walls of limestone, where the channel is only 150 yards wide. Its deficiency in space is made up in the violence of the current, which runs with inconceivable swiftness, the waters rushing down a frightful rapid called the "Falls."[7]

The Miramichi [which in the Micmac Indian language

signifies 'Happy Retreat'] is the second river in extent and importance of the province. Its branches, which are very numerous, drain a vast tract of wilderness country, and, by being united as they approach the sea, they form a stream of considerable magnitude. Some of its north-western branches approach the St. John, and almost touch the Nashwaak, others reach the lands of the lower Tobique. Three of the north-west branches spring from a chain of lakes in the upper Tobique country. Having descended with considerable rapidity from its principal source and traversed the forests of the south-west nearly 200 miles, the branches of the Miramichi unite and become navigable for large ships; and finally, the river makes its *débouchment* into a spacious bay of the same name ...

The banks of the main stream are settled 100 miles from the bay, and the mouths of some of the principal branches are also thinly inhabited; but remote from the larger tributaries, the country is in the original wilderness state, and millions of acres of land capable of successful cultivation are covered by dense forests, and even the fine tracts of intervale on the borders of the streams to a great extent remain uncleared.[8]

Along the banks of the river, there are numerous wharves and landing-places, depots for timber; but the principal business of the country is carried on at Chatham, Douglas, Newcastle, and Nelson — four small towns situated within a distance of five miles. The extensive district bordering upon the Miramichi and its tributaries has derived its chief importance from the vast quantities of valuable red and white pine that formerly stood upon its lands; but in 1825 this part of New Brunswick was visited by a most awful and calamitous fire that consumed the forests like stubble, and besides destroying a number of the inhabitants, involved the whole population in ruin and distress. From the great annual exports also, the timber is growing scarce, and more difficult to be obtained; so that the period is fast approaching when the great number of persons employed in lumbering will engage in the more permanently profitable occupation of husbandry.[9]

The principal sources of the Restigouche are situated in a mountain range that extends through the whole District of Gaspé. The course of the river, from its mouth to the distance of 60 miles, ascending, is to the south-west; it then turns at a right angle to the north-west ...

The Bay Chaleurs, having extended deeply into the country, finally terminates in this fine river, which opens a wide district to all the advantages of trade and internal communication. The banks of the Restigouche are not settled more than 30 miles above its mouth. The upper part of the noble stream, and all its branches, pass through a dense wilderness, and the whole interior of the country is uninhabitable.[10]

The Petitcodiac takes its rise near the sources of the Kenebecasis, and having run in a north-easterly direction 40 miles, turns at a right angle, called the Bend. It then runs to the south 20 miles, and discharges its waters into Shepody Bay. It is navigable for vessels of a 100 tons burthen 30 miles from its mouth, and large ships are laden at its curve. Here the tide flows in and ebbs off in six hours, and runs at the rate of seven miles an hour. The flood-tide is accompanied by a splendid "bore," or tidal wave, which at spring-tides is five and sometimes six feet high. The rushing of this overwhelming wave is accompanied by a noise like distant thunder, and affords an interesting spectacle ...

The Memramcook, Tantamarre [now Tantramar], Aulac, and Missiquash are small rivers which, like the Petitcodiac, pass through very extensive marshes.

The Great Tantamarre Marsh is situated on both sides of the river of that name. It is about 12 miles long, and, upon an average, four miles wide, being one of the most extensive collections of alluvium formed by the sea in America. In the Parishes of Sackville, Dorchester, and Moncton, 4,900 acres of marsh have been rescued from the sea by dikes and embankments. All the streams emptying themselves into Shepody and Cumberland Bays are skirted with alluvial deposits, which are more productive than any other lands in the country ...[11]

The whole of the southern coast of the Province, from the River St. Croix to the entrance of Chignecto Bay, is bold and rocky; it is, nevertheless, indented by a number of fine harbours, and small deep bays, in which the largest ships may ride in safety ...

The whole of Passamaquoddy Bay is studded with islands, which are said to be 365 in number. Many of these islands are merely rude masses of rock, or small eminences covered with moss and stunted spruce. The larger ones have a soil of medium quality, and produce excellent potatoes, barley, and oats. During the summer or fishing season, the bay presents an interesting spectacle. Boats and vessels becalmed are swept away by the rapid tide. At one instant they are hidden by some blackened rock, at another they are seen gliding from behind the green foliage of some little island. Sometimes hundreds of boats are huddled together, their crews being actively engaged in drawing forth the finny inhabitants of the deep. As soon as the shoal of fish retreats, or the tide is unfavourable, they disperse, and the surface of the water is decorated with their white and red sails. The Indian, in his frail bark canoe, without a rope or anchor, is also there, and the report of his gun, discharged at the rising porpoise, is re-echoed among the cliffs of the shore. Flights of gulls hang over the glassy surface of the water, which is here and there broken by a shoal of herring, or the spouting grampus in search of his prey.[12]

3
Descriptions of Some Favorite Places in Prince Edward Island[1]

The geological survey of Prince Edward Island has been completed, and if the province has not been favoured by Providence by any very rich deposits of fuel, or the metals, it is presumed that the benefits conferred upon its agriculture, will amply repay the small sum expended in the undertaking.

The chief part of the Island is beautifully variegated with hill and valley, and numerous small bays, rivers, and creeks, lakes and lagoons greatly contribute to the beauty of the scenery, which, although not lofty and majestic, is peculiarly interesting. The entire surface is abundantly supplied with springs and rivulets of the purest fresh water. Descending from the more elevated ridges of land, numerous streams fall in opposite directions, and although in a low country, these afford less power to propel machinery than in higher districts, they are extensively employed in working flour and saw mills, carding mills, &c. The mouths of almost all the rivers and creeks are skirted by small tracts of salt marsh, deposits of marine alluvium, shells and plants. Along the eastern shore of the island there are extensive collections of drift and blown sand. These moveable deposits are often thrown up into picturesque mounds, and by being stretched across the mouths of the bays and rivers, they form safe harbours and tranquil lagoons ...

Peat bogs are very numerous, although few of them are of any great extent. The largest and most valuable deposit of peat on the island is on the south side of Cascumpec Harbour. It contains a buried forest, and, as the quality of the peat is very superior, it will, in the course of time, be valuable for fuel ...

The East, North, and West Rivers were explored by the aid of Indians and a large canoe. The shores are seldom bounded by cliffs, but descend gradually to the water, being frequently skirted by tracts of peaty ground, salt marsh, and a mixed alluvium ... Reposing directly upon the rocks there are frequently thick deposits of clay. One of these occurs opposite the town, near the Ferry Wharf, and on the property of Mrs. Desbrisay, and is very favourably situated for an extensive manufacture of bricks ...

Viewed from the signal station, or either of the old French forts at the entrance of the Harbour, Charlottetown, and its surrounding scenery, are very beautiful, the shores, in every direction, are cultivated, and tracts of native forest are interspersed with fine fields and spacious farm-houses, with these, a number of ships upon the stocks afford a peculiar contrast ... The buildings in general are more in English style than is always seen in British American, and wide streets and open squares contribute much to the comfort and health of the inhabitants ...

Rollo and Colville Bays are convenient harbours and the populous villages of the French add much to the beauty of the natural scenery ... This shore [Colville Bay] was evidently inhabited in former days by the native Indians, and, from the character of their relics, they appear to have been Micmacs, the descendants of whom are still upon the island. These relics consist of axes, spears and arrow points, and rude pots made of stone — barbed fish bones, which they employed in fishing are also found. Some of the arrow heads are made of Labrador feldspar, agates, hornstone and jaspar. The feldspar is identical with that found at Labrador; the agates are like those of the Bay of Fundy, and as none of these minerals have been found *in situ* on the island it is very probable that the pieces used by the Indians were brought from those places ...

St. Peter's Bay is a narrow but deep indentation, and a safe harbour. Its mouth is protected by a chain of sandhills, having a narrow channel between them that is capable of admitting large ships at certain times of tides. These sandhills resemble the cones of extinct volcanoes: they are liable to constant change, and were they not covered with bent grass, they would be much more liable to drift away before the winds than they are at present. Near the mouth of the bay, a forest of hardwood has been buried by the drifting sands: the ancient channel of the river has been filled up; and the wharves built by the French, who were the first civilized inhabitants, have all been buried in the shifting shingle. An opening formed by the sea during a gale, exposed a thick bed of oyster shells and a number of Indian relics.

The scenery of this Bay, with the surrounding country and its fine farms, is very beautiful ... This Bay has afforded one of the best salmon fisheries of the Island ...

Between the head of Hillsborough River and Savage Harbour there is a tract of low land, across which, it is probable the tide once passed between the eastern and western coasts. Savage Harbour has a narrow and shallow inlet, situated between low sandhills ... By the

Micmac wigwam, Prince Edward Island Archives and Record Office, No. 3109/14.

Rural Prince Edward Island, Prince Edward Island Museum and Heritage Foundation Collection, Prince Edward Island Public Archives and Records Office. No. 3466/H.F. 7266.3.3.

encroachments of the sea on the south side of the harbour, a number of Indian skeletons have been exposed and washed from the bank. These skeletons were lying together in different positions, as if the bodies had been thrown into a common pit, the top of which was only one foot beneath the soil. From an examination made at the spot, some of the bones were found to be of great size; and in general they all exceeded in their dimensions those of the race in its present state. The site of this pit, on the extremity of a small point of land, supports the opinion that the savages had been surprised and cut off, or killed in battle, and as no relics of warlike instruments were found at the place, except those of the aborigines, it is probable that the event took place before the island was inhabited by Europeans. From an old tradition of the affair among the Indians, the bay has been called Savage Harbour ...

At Tignish, near the Cape [Cape Kildare], there is a large village of Acadian French, and two fishing establishments. Shore fishing is carried on by some of the inhabitants.

Formerly this place and the cape were the resort of great numbers of the Walrus or sea-cow: hundreds of those animals were killed on the land by the early inhabitants, among whose descendants pieces of their skins still remain in use. A deep pond near Tignish is said to be filled with their bones, and their tusks of ivory are occasionally found on the shore, or in the forests. Only a few of those noble animals are now seen, and of their number, which is stated by the fishermen to be on the increase, none are [sic] captured...

The main road passes through the Miscouche Settlement, a large village of French Acadians. At the point there are also upwards of 50 families of these frugal and orderly people ...

Returning to Bedeque, a great change is observed in the

general features of the country. The lands are more elevated, and the surface is diversified by hill and valley. The soil is extensively cultivated and produces excellent crops of grain, and all the vegetables and fruits of the climate. The scenery is revived, and a view from the fine farm of Capt. Thomas, or from any part of the banks of Dunk River, is very beautiful. The mouths of the rivers, celebrated for their fine oysters, are skirted by tracts of salt marsh and marine alluvium, abounding in shells, which, with the limestone that may be collected at different localities, offer abundant resources of manure ...

The rocks of the Island agree in their lithological characters with those of the opposite coast of New Brunswick, where they form the shores from point Escuminac to Bay Verte; and the physical geography of the country corresponds with that of a large tract bordering upon the gulf in the counties of Kent, Westmorland and Miramichi. The inclination of the strata, also, have a general agreement on both sides of Northumberland Straits, in which the water is shallow. A dangerous reef of Cape Tormentine extends, in the direction of the strata, towards Cape Traverse. It is therefore not improbable that at some former period, the island was separated from the main by the breaking through of an isthmus that united them. How far such a result has been promoted by the powerful currents of the Gulf of St. Lawrence, it would be difficult to determine without an accurate knowledge of their direction and an estimate of their forces ...

On a part of the north side of the island, where the coast is exposed to gales that sweep across the gulf, the shores, after having been greatly intruded upon, are bounded by chains of sandhills. Near the North Cape, between St. Peter's Bay and East Point; between the North Cape and West Cape, and at other places on the southern extremity of the island, the sea is still making rapid encroachments, and is annually reducing the area of Prince Edward Island. Even in some of the bays and harbours this encroachment is so rapid that the cemeteries of the dead have been broken into and the mortal remains of their tenants have been washed away by the waves.

The hard rocks of Point Prim have resisted the advance of the sea, while the clayey and friable strata of Orwell Bay are yielding to its sway. It is certain mineral matter thus removed is again thrown back upon the coasts, still the loss of the dry land far exceeds the accumulations of sand and alluvium lodged in the bays and upon the shores. It would be difficult to estimate the annual diminuation of the island from the above causes. It is, however, very considerable, and far beyond prevention by human means.

4
The Micmacs of Nova Scotia[1]

Census of the Indians of Nova-Scotia Proper, 25th September, 1847

Counties	Males	Females	Totals
Annapolis	33	33	66
Digby	46	58	104
Yarmouth	32	24	56
Shelburne	23	24	47
Queen's	66	57	123
Lunenburg	18	14	32
King's	24	19	43
Cumberland	16	15	31
Pictou	63	59	122
Sydney	61	58	119
Hants	54	57	111
Halifax	58	49	107*
Totals	494	467	961

Indians in Nova-Scotia Proper	961
Indians in Cape-Breton	500
Total of Nova-Scotia	1461

[Micmacs of New Brunswick, in 1841   935
Micmacs of Gaspé, 1841   411
Micmacs of Prince Edward's Island, 1847 250
Micmacs of St. Pierre, Newfoundland, 1847 200]
   Total number of the Tribe   3290

The average number of children to each family is now only 2 1/2.

Among the 961 of Nova-Scotia Proper, there were in 1846: births, 79; deaths, 106. At this rate of decrease the whole race will be extinct in 36 years, and that result is certain unless measures are immediately adopted to prevent it.

*The Indians who fish at Canseau [Canso] belong to Cape Breton.

Indian Scene, W.H. Bartlett, J.C. Armytage, **Canadian Scenery** (Volume II), 1842.

Melancholy, indeed, is the reflection arising from these details, and still more painful is the consideration that the destruction of the whole Micmac race, the ejected owners of the country, is still advancing with fearful rapidity, for they are falling like the leaves of their native forests before the withering autumnal frost. If her Majesty's and the Provincial Governments desire to save the small remnant of the Aborigines, the work of justice and humanity must be commenced immediately. Unless the vices and diseases of civilization are speedily arrested, the Indians of these provinces will soon be as the red men of Newfoundland, or other tribes of the west, whose existence is forever blotted out from the face of the earth. If, like the beaver of the lakes they are to become extinct, may the page that registers the event to posterity also record the noble efforts of a Christian nation — a Christian province — to smooth the path trodden by the devoted race.

The efforts of the Micmacs to resist the invaders of their lands and liberties were just and natural. Without religion or civilization they practised their peculiar mode of warfare, and its barbarities were increased by the merciless and wanton cruelties of the early European *Voyageurs*. They were exposed to the most vindictive ferocity. In Nova Scotia, the Soldiers were ordered to spare the disaffected Acadians, but to give the Indians no quarter — they were hunted like the wild animals of the forest, and the 'Caughnawaga' (the place where Christians lived) was to them the place of destruction.

On the 1st of July, 1761, and after Acadia or Nova-Scotia had been conquered and secured to the British Crown, a Treaty was entered into with the Indians, and Argimautt, the Chief, ratified that Treaty at Halifax with great ceremony. Having obtained every submission from the Chief, 'the Commander in Chief took Argimautt by the hand, in token that His Majesty received him into his favor and protection.' That favor and protection is what these people still claim. Since the treaty, the Aborigines have always considered themselves under the immediate care of the Crown; but their hopes and the promises held out to them when they rendered their submission, have never been realized. The gifts made to them from time to time have increased their indulgence in idleness and vagrancy, and no improvement whatever has been made in their condition. For the lands, forests, and fisheries, long since taken from them, they are of the opinion the Government should make a far greater compensation than they have ever received, or the permanent protection contemplated by the Chief at the time when they laid down their arms and 'smoked the pipe of peace.'

The Micmacs were never rebels nor traitors: they are the original inhabitants of a conquered country. Nearly 90 years have passed away since they became British subjects. In that period nothing has been done to civilize the race, now brought to the lowest depth of misery and despair.

Diminuation of number and the final extinction of a savage race, yielding their territory up to civilized occupants, is a feature not peculiar to America. It might be supposed that after their mutual wars had ceased, and their encounters with the whites had terminated, the Aborigines would multiply, yet experience his proved exactly the reverse. Among the most prominent causes of the decrease of the natives has been the introduction of European diseases. Exposed to the inclemency of the weather, and destitute of the proper diet and treatment required by contagious diseases, numbers are swept off annually by complaints unknown to them in their original state.

During my tour of inspection, I prescribed for several cases of hopeless consumption. The venereal disease, the scourge of vice contracted by the visits of the dissolute to the towns, is by no means rare, and I have seen its direful effects on children at the breast. Pulmonary consumption is frequently induced by intoxication and exposure to severe cold. The law prohibiting the sale of spirituous liquors to the Indians is almost disregarded, and thus the morals and health of these people are undermined. Of late many have taken the Temperance Pledge, and general sobriety, with increased industry, begin to prevail.

The erection of dams across the rivers has destroyed some of the best salmon and alewife fisheries in the province. The best shore fisheries are occupied by the white inhabitants, from which the Indian is sometimes privey by force. From the clearing and occupation of the forests, the wild domain of the Moose and Caribou has been narrowed. Being hunted by the dogs of the back settlers, these animals have become scarce — thus the Indian has been deprived of his principal subsistence, as well as the warm furs that in olden times lined his wigwam. Indigenous roots once highly prized for food have been destroyed by domestic animals. Herds of swine have consumed the shell-fish upon the shores. To these may be added the actual driving back of the Indian into the interior woods, whither the food, obtained by the prices of their baskets, or by begging, must be carried upon their backs. These united causes have operated fearfully, and at last reduced the whole tribe to the extreme of misery and wretchedness.

The Micmacs are naturally full of humour, and very contented. They are now a discouraged and spirit-broken people. In many instances the lands reserved for, and occupied by them, have been encroached upon — they have been forcibly driven away from their old camping grounds, and therefore, a number of families are seen wandering over the country, often almost destitute of decent clothing or a *chantier* to shelter them from the storm. Many have perished. In their primitive state, the government of these people was patriarchal. Due restraint was put upon society by their Chiefs or Fathers. Buying and selling were unknown. Every transfer of property was a bona fide gift. Their negligence in the payment of debts at the present time arises from that custom. They worshipped *Kesoult* the great spirit, in which they still believe, rather than the doctrine of the trinity. They were kind to the aged and infirm, and their priests administered decoctions of the roots to the sick under strange incantations. Their attachment to their tribe, their patriotism for its honour and welfare, were not exceeded by the Greeks or Romans. They had their priests, their chiefs, their councillors, orators and warriors. Their wants were few, and they were happy. Painful indeed has been their lot since they were conquered — conquered 'not to redeem, but to destroy.' The scattered remnant of this once brave and patriotic people is now utterly degraded and overwhelmed in misery. They have been supplanted by civilized inhabitants, and in return for the lands of which they were the rightful owners, they have received loathsome diseases, alcoholic drinks, destruction of their game, and threatened extermination. More than once have I seen the tears trickle down the furrowed cheeks of aged Indians as they recounted the losses of their Tribe by what they always call an impolitic Treaty.

Almost the whole Micmac population are now vagrants, who wander from place to place, and door to door, seeking alms. The aged and infirm are supplied with written briefs upon which they place much reliance. They are clad in filthy rags. Necessity often compels them to consume putrid and unwholesome food. The offal of the slaughter-house is their portion. Their camps or wigwams are seldom comfortable, and in winter, at places where they are not permitted to cut wood, they suffer from the cold. The sufferings of the sick and infirm surpass description, and from the lack of a humble degree of accommodation, almost every case of disease proves fatal. In almost every encampment are seen the crippled, the deaf, the blind, the helpless orphan, with individuals lingering in consumption, which spares neither young nor old. During my inquiries into the actual state of these people in June last, I found four orphan children who were unable to rise for the want of food — whole families were subsisting upon wild roots and eels, and the withered features of others told too plainly to be misunderstood that they had nearly approached starvation. In one camp I found one lunatic, one blind, one deaf man, with two persons, each over 90 years of age. Scenes equally unpleasant are not rare.

The Indians display much skill and ingenuity, and they are quite equal to the whites in natural understanding and ability. Their powers of endurance and patience under the greatest trials are truly remarkable. They are still disposed to change their places of abode. In winter they erect their wigwams in some sheltered wood — in June they move to the shores and fishing grounds — in autumn they retire to their favorite retreats. Some families are constantly in motion, and in the space of a few hours, a whole encampment will disappear, and nothing remains at the site but the naked poles of their tents, and the ashes of their camp fires.

The Micmacs have a written language which consists of hieroglyphics, resembling Chinese characters. Manuscript books of prayers, and portions of scripture, introduced by the Jesuit priests, are common among them. Their sacred music is also written. A few individuals can read English, and several children have been taught in the common schools of the country. Since their submission to the Crown, they have, as a body, retrograded, and the present generation is extremely ignorant. They display excellent capacities for learning, and are submissive to discipline. Their minds are strong, and their powers of imagination extremely fertile. Their language is full, bold and figurative, and their words are readily multiplied according to the ideas required to be expressed.

The Government has very judiciously reserved tracts of land in different parts of the province for the Indians; they are in the

	Acres
County of Halifax	1300
County of Hants	1750
County of Cumberland	1000
County of Lunenburg	2000
County of Queen's	1000
County of Digby	1000
County of Sydney	1000
County of Annapolis	1000
County of Cape-Breton	12000
Total	22050

800 acres at Margaret's Bay, and a lot at the mouth of River Philip, formerly reserved, have been alienated. My predecessor, Mr. Howe, had a lot of 1000 acres surveyed for the Indians at 'Kedgum Coogie' [now in Kejimkujik National Park] or Fairy Lake, in the County of Liverpool, and upon which he succeeded in settling several industrious families. It is desirable that this lot should be reserved with as little delay as possible. The reserve in Cumberland, and 1000 acres in the County of Lunenburg,

have never been surveyed. A few families have commenced clearings upon ungranted lands at Argyle. A reserve of 1000 acres, at a proper situation in this quarter, would probably collect the wandering families in this district. It is necessary that the lines of the Indian lands at Gold River should be run and marked, to prevent altercation that has already arisen upon them. A small tract upon the Cornwallis River, and another in the Gaspereau Lake, in King's County, have been resorted to by the Indians for centuries past — on them are burying grounds and small orchards. These lands, unfortunately, were never reserved, and they cannot be obtained without purchase or expensive lawsuits. The Indians have been deprived of the lands they formerly occupied at many other places. This is much to be regretted on account of the injustice done to them and the discouragement they have fallen into on that account.

The quantity of land reserved for the Micmacs of Nova-Scotia Proper is about 10 acres for each soul. At the present period this is a sufficient supply, especially if the Tribe shall be allowed to become extinct. Some of these lands are almost worthless for agriculture and are very unfavorably situated. The propriety of setting apart for the uses of the tribe a few more tracts at proper sites is worthy of consideration.

Trespasses are committed upon the Indian reserves with the most daring impunity. I have made efforts to check the removal of timber from these lands; but the remoteness of their situations renders the task almost unavailing. As the soil must be the foundation of every improvement and the civilization of the Tribe, it is necessary that their lands, and the timber upon them, should be carefully protected.

In reference to education, little progress has been made during the past year. The settlement of these people in villages, and the reclaiming them from a wandering life, must precede the establishment of schools. Such as now reside within the reach of tuition may be instructed; but the very limited means placed at my disposal, in consequence of the pressing necessities for food, have not permitted me to embark in the education of the young.

In all my communications with the Micmacs, I have received the kindest marks of their attention. From a steady acquaintance I have had with them for a number of years, and the observations I have made during the past season, I am convinced that by judicious management and active measures, they may be brought to a state of perfect civilization, without which they will finally disappear, and nothing will remain of them, save the unhappy remembrance of their fate.

They are now deprived of almost all their original sources of maintenance, and compelled, by absolute necessity, to adopt the habits and industry of the Colonists, without which they cannot subsist. It has been maintained that christianization must precede civilization — if so, the first work has in some degree been completed. No attempt has ever been made to bring the wandering Micmac from his wild pursuits to sober industry. Whatever has been gained towards this point has been accomplished by the unaided efforts of the Indians themselves, directed by the power of imitation and necessity. That their conversion to Christianity has tended to this result is by no means certain. At the rate of decrease these people are now advancing, the day of their arrival at civilization will be that of their final extinction.

We have here a people of active capacity, having a voluminous written language, and who are well known to acquire learning with great facility, already half civilized by their own efforts. Surely this desirable work can be completed — and the time has arrived when justice and philanthropy call loudly for its accomplishment.

The only way that the bounty of the government can be of any benefit to the Indians, is by applying it to objects that will finally enable them to take care of themselves. It is their permanent, and not their temporary relief that is required. The sums annually applied in donations, and purchase of blankets, etc., for the Aborigines, are lost to the highest objects of their welfare; this fact is acknowledged by their chiefs and captains. In some instances, they produce jealousies and contentions. In others the gift is wasted in intemperance, and it is only in extreme cases, under very judicious management, they at all alleviate suffering. The whole sum granted by the legislature [of Nova Scotia] at present, for the benefit of the Indians, amounts to 4s. 1d. per head, and every individual claims a share in the donation. The consideration that now presents itself is of deep importance — namely, whether these people are to become extinct, and their last descendant disappear for ever, or by increased exertions on the part of their fellow men, they are to be redeemed from the destruction that awaits them, and brought to enjoy all the blessings of their colonial brethren.

5
Native People of New Brunswick

At the present time, there are the remnants of two tribes in New Brunswick — the Micmacs and the Melicetes, or Morrisetes. The former are found in a part of the District of Gaspé, on the whole coast from the Restigouche to Bay Verte, and on the entire surface of Nova Scotia. The latter reside chiefly along the valley of the St. John, on the banks of the St. Croix, and the country westward, where they are met by the Penobscots of the United States. The Micmacs speak a dialect of the Iroquois (or language of the Six Nations), Hurons, and other tribes of the North: but the Melicetes, from being descended from the Delaware stock, speak a dialect of that people which is scarcely understood by the descendants of the Iroquois.

The physical characteristics of those people are not dissimilar. They all have the same copper colour, the straight coarse black hair, hazel eyes, high cheek-bones, scanty beard and erect carriage, common to all the northern tribes. Some of the men are upwards of six feet in height, and remarkable for suppleness, activity and great powers of endurance, rather than for strength. Individuals among them will travel 70 miles in a day without any apparent fatigue. Such feats are often performed under heavy burdens, and without any kind of food. Bears, deer, moose, and other wild animals are sometimes pursued by them and overtaken. The skill and agility they display in ascending and descending the dangerous rapids on many of the rivers in their canoes has never been attained by Europeans; and the quickness of their perceptions in discovering the trails and footsteps, and even the scent, of men and animals, is truly surprising.[1]

From the migratory disposition of these people, they seldom erect comfortable habitations. Their huts, called *wigwams*, consist of ... poles, set up from the ground and covered with the bark of trees, except at the top, which is left open to allow the smoke to escape; when completed, the whole fabric resembles a cone ... A village of these singular dwellings, situated on the bank of a river, and sheltered beneath a grove of lofty trees, is almost the climax of American scenery.

The Indians display more skill and workmanship upon their canoes than on any other fabric. The frame consists of the strongest and lightest kinds of wood, bent in their hoops to the approved model: on this is laid the entire bark of a large white birch tree, which is often found of sufficient size to cover the whole exterior of the vessel, and being cut and fitted to the framework, is finally sewed at the ends and to the gunwale with the tough roots of the spruce; the bars, or cross-pieces are also sewed in. These vessels are not only well adapted for shallow and rapid streams, but, from their lightness and buoyancy, they are safe amidst the stormy waves and breakers ...

The model of the canoe of the Melicete differs from that of the Micmac. It is somewhat in the style of an English barque, and has a very graceful appearance upon the water; but the canoe of the Micmac is rather the safer vessel in an open sea in stormy weather ...[2]

The social condition of these people was like that of the Canadian tribes. The independence of every individual to do whatever he pleased was maintained as a right, and that principle has not been weakened by their intercourse with European communities. They will enter the dwellings of the rich and sit at the tables of persons of rank with an air of dignity and self-possession, and their powers of imitation remove every kind of awkwardness. Their attachment to their tribe, and their patriotism for its honour and welfare, were not exceeded by the Greeks or Romans ...[3]

The acuteness of the Indian is almost supernatural; he can follow an animal by indications imperceptible even to an American backwoodsman. His powers of observation are so perfect that he can trace on a piece of bark, with a bit of charcoal, the geography of the country he has traversed; and he will take a direct course to a place hundreds of miles distant, without the aid of a compass.

It has been supposed by some writers that the savage tribes of North America had no means of recording events. The wampum belt was generally applied to the different parts of a speech, or the different articles of a treaty; and on great occasions, when these belts were brought forth, individuals were found who, from memory or tradition, could explain each section of the precious girdle: but, besides this mode of record, the Micmacs and Melicetes had pictorial representations of certain events, and communicated information through the medium of hieroglyphics. Rocks and trees in conspicuous situations have had figures cut or engraved upon them which convey to the Indian traveller in concise terms the knowledge necessary for his safety and comfort.

During his geological survey of the province, the writer, with two companions and three Indians, were much embarrassed in not being able to discover in the wilderness an old Indian portage between the head waters of the St. Croix and Eel River Lake. From this difficulty they were relieved by observing some rude hieroglyphics marked upon an old cedar tree. The representations were that of an Indian carrying a canoe, and the direction of the figures corresponding exactly with that of the portage path, which had been obscured by grass and fallen leaves. A hunter with his gun levelled at two deer, indicated that those animals were plentiful: this, and other information conveyed in a similar manner, was found to be correct.

In another instance, when the same party was descending Eel River, and their lives were in jeopardy on the brink of a fall, a large drawing of two Indians, with their heels uppermost and their canoes capsized, was seen executed in durable black ink upon a broad piece of cedar secured to a post: this warning was immediately understood, and a landing was effected before the canoes and the whole party were plunged down the cataract ...[4]

The great number of lakes and rivers in the province afforded the Aborigines great facilities of water communication. The distances between those lakes and the sources of the rivers by land are short, and are called portages, which in the forests, are only narrow and obscure paths. On some of the ancient Indian trails the solid rocks have been worn out by the moccasins of the natives tribes ...[5]

In 1841, Sir William Colebrooke, the Lieutenant-Governor of the province, appointed M.H. Perley, Esq. a special commissioner for Indian Affairs, by whose labours much information was obtained of the condition of the two tribes, and the state of their lands. According to the report of the commissioner, the number of Melicetes, male and female, is 442; of Micmacs, 935; total number of Indians in the province, 1,377. There was a time when those tribes could muster more than four times that number of warriors, exclusive of other population; and they now believe that the diseases of the Europeans, and ardent spirits, have been more fatal to them than the arrows and scalping-knives of all their enemies.[6]

6
Forests of Nova Scotia and New Brunswick
With Pronouncements on Abundance and Varieties of Trees, Fires, Succession of Forest Trees and Exploitation

The chief part of the wild lands is covered with the native forest, and trees in every stage of growth and decay. To know the forest is to live in the forest. Such as would gain a knowledge of its beauty or gloom must visit it at different seasons of the year, he must sleep upon the mosses in summer, when the earth is shaded by a living mantle, in whose folds the feathered songsters pipe forth their shrill melodies — he must see it in its gay autumnal dress of every color, and also in winter when all but the Pine tribe stand naked and leafless, bending and creaking before the cold northern breeze.

It is in these wild interior districts that the moose and caribou still roam at large, and the bear stalks forth in search of prey. The forests of Nova Scotia are not surpassed in beauty by those of any other part of North America. It is true that all the best timber that grew adjacent to the bays and rivers, has long since been felled and removed. The supplies now sent to the British market, are obtained by great labour, and therefore the timber trade of the province is below that of either Canada or New Brunswick.[1]

Every country has its epochs of industry: the present, in New Brunswick, is the timber period, which will be followed by the agricultural, fishing, and, finally, the manufacturing eras...[2] Sixty years ago, almost the entire surface of New Brunswick was an unbroken wood, and the first settlers carried a musket in one hand to protect themselves from the assaults of the Indians, and an axe in the other to clear away the trees...[3]

Forty years ago, the resources of this part of the province (County of Northumberland) were almost unknown, except along the seaboard. The banks of the main stream, and those of its branches, were found to abound in groves of red and white pine, of which very extensive shipments have been made to the mother country. Much of the valuable timber has been removed, and a still greater quantity was destroyed by an awful conflagration in 1825...[4]

The forests of pine, spruce, fir, and other resinous woods in America, are very liable to be consumed by fire, and there is scarcely any part of the northern continent that has escaped conflagration, either before or since the continent was taken possession of by a civilized people. The different tribes of Indians have traditions of fires even more destructive than the one it is now our painful duty to record; and the charred forests buried in peat-bogs or alluviums bear testimony to the occurrence of those vast surface burnings, that are more to be dreaded than the floods of liquid lava that buried the cities of Herculaneum and Pompeii. The great heat and frequently dryness of the summer months cause the resinous juices to exude from balsamic plants, whereby they are rendered very inflammable; and were it not for the almost universal admixture of hard woods, whose juicy leaves resist the spreading of the lashing flame, such catastrophes would be more common, and far more general in their operations.

In every American state and colony, laws have been enacted to prevent the firing of the woods in the summer and autumn; but where the clearing of land by burning the timber is an essential occupation, it cannot be surprising that accidents will happen, or by carelessness the devouring element gains the mastery, and rushes forward with devouring energy. Again, dry trees are sometimes fired by lightning; and the fire remains unextinguished until the moisture of the passing shower is dried, when it spreads abroad, being fanned into fury by the summer breeze...[5]

When the trees of the forest have been destroyed by fire or cut down by the axeman, and the land neglected afterwards, of whatever kinds they were, other trees succeed them. Land cleared of its lofty pines and spruces is soon covered by poplars, maples, larch, fir, wild cherry, cedar, and whortleberries. Sugar maple, beech and larches are frequently succeeded by red pine, spruce, raspberry and gooseberry bushes. Hemlock land is soon overspread by alder and maple. Oak is followed by fir and spruce. The original kinds of trees seldom appear in the second growth; but after the second growth has been removed, the first, or one differing from both, occupies the ground. Lands that formerly bore yellow birch and hemlock are now bearing sapling oaks and poplar. Some of the old pine lands that have been under cultivation nearly a century, if laid down to pasture, will soon be covered with shrubby birch and poplars. This rotation seems to be a law of nature, and intimates to the farmer the necessity of changing his crops. In general the evergreens are succeeded by the hard woods...

Much stress has been laid on the discovery of the qualities of the soil by the trees growing upon it. Some trees are found most abundant and thrifty in certain districts; but this depends more upon the dryness or moisture of the soil than upon its peculiar properties, as the hard woods are succeeded by the evergreens, and vice versa, it is quite impossible to determine the character of the soil by the presence of either, nor would the difficulty be diminished if the history of the ancient forest were well understood. There are frequent instances where the surface once occupied by large beeches, birches and maples is finally overrun with laurel and whortleberry bushes which have been supposed to indicate extreme sterility. Farmers have been agreeably disappointed in finding such lands fertile. I have seen fine wheat, barley and oats grown upon laurel and fern land, and patches almost barren are sometimes studded with close groves of hardwood saplings...[6]

White pine ... is the richest production of the forests of British America. As a staple article of commerce and domestic use, it is unrivalled. It is the largest and most valuable tree eastward of the Rocky Mountains; frequently it rises to the height of 200 feet, with a trunk five feet in diameter. I have measured pine trees on the late disputed territory between New Brunswick and the United States that were 18 feet in circumference. The lofty pine towers far above his indigenous associates and is leafless except at the very summit. It is the monarch of the wilderness, which seems to aspire to reach the clouds, yet it bends before the gale, and waves its umbrella-shaped top high in the air. When it is felled, it crushes down the smaller wood, and by striking uneven ground, it is sometimes broken. This noble tree flourishes best on deep sandy soil, although it will not refuse to grow on ridges of granite and amidst shapeless rocky boulders. The wood is straight, fine, light, free from knots, and most easily worked; hence it is almost universally employed. It supplies masts for the largest ships, square timber, deals, boards, scantling, clapboards, shingles, and the wood used for the finishing of every kind of carpenter's work. Its applicability to the common requirements of modern architecture have rendered it by far the most valuable of all the native trees...[7]

Lumbering, as it is called, still remains to be noticed. In the early settlement of the province, the districts now covered with fine cultivated fields, and the banks of the rivers and minor streams, were covered with different members of the pine tribe, or groves of marketable hard wood. Most of these have been felled, and consumed by fire in clearing the land, or they have been shipped to Great Britain and the West Indies. Timber is now obtained at far greater expense than formerly, and the prices received for it do not remunerate the lumbermen for their privations, labor, and risk. In the best of times, the actual hewer or sawer of the wood was not the gainer, and the fluctuations in the demand and prices have allowed but few timber merchants to enrich themselves, while bankruptcies have been frequent among them.

The timber trade has drawn the labor away from agriculture. Thousands of farms have been abandoned or neglected. By it, habits of unsteadiness and intemperance have been induced among a part of the laboring population; and up to the present time, many families are depending upon the falling of showers of rain to float the wood rolled into the streams, or to set the saw mill in motion, whereby they obtain their subsistence, rather than from the produce of the soil. In the neighbourhood of sea ports, fine farms have been deserted and turned out to common; and instead of stacks of hay and grain, the dilapidated buildings are surrounded by piles of the shavings of shingles and hoop poles, the result of the means by which the proprietors obtain a scanty supply of the necessaries of life. Who are the most independent men in the province? Who are they that live most comfortably, and enjoy the highest degree of civilization? They are the farmers — a class of men who never meddled with either lumbering or fishing, but devote all their energies to the tillage of the soil.[8]

7
An Initial Profusion of Game and the "Wanton and Wasteful" Destruction Of This Resource

When British settlers first began to immigrate into the province, game of all kinds was plenty. The rivers and lakes were alive with ducks and teal, the woods with partridges, and the marshes with plover. Moose were killed in great numbers, and venison was the only kind of flesh consumed by many families. There was then a wanton and wasteful destruction of the giants of the forest. Thousands were shot down for the sake of their skins; and up to the present time the flesh is often left to putrify, or to be devoured by the carnivorous races.

The practice of hunting the elk with packs of dogs, when the snow is deep, is very destructive. The Indians have petitioned the Legislature for the enactment of a law to prevent this mode of hunting the moose, by which they will soon disappear altogether, and the Micmacs be deprived of the food they have always considered to be their own. Without a protecting law, assailed at all points and at every season of the year, it is remarkable that any of these noble animals have survived; but they are still to be found in all the remote uninhabited districts, and frequently they not only approach the settlements, but emerge from the woods into the cultivated fields, and during the rutting season they frequently cross from one line of forest to another, regardless of every danger. Notwithstanding their fleetness, in like manner have the reindeer, or caribou been destroyed...

In winter these animals form what is called a yard, which is merely a tract of ground upon which they feed and beat down the deep snow while browsing upon the evergreen herbage. In such a yard they will remain all winter if not disturbed. Their great weight prevents them from travelling over deep snows, and when they are covered by a crust of ice the animal seldom escapes from a persevering chase.

With moccasins and snow shoes, the hunter enters the yard and shoots the harmless herd. If any attempt to escape, they are followed and cut down with the tomahawk, or axe. This is a barbarous amusement or savage butchery from which a true sportsman would turn away in disgust. Indeed the gallantry of the work scarcely exceeds that of entering a farm yard and slaying a drove of cows...

The moose is not a fleet animal; he has a shambling gait like an ox. During the chase he will frequently turn upon the dogs, take them up on his horns, and toss them into the air. He frequently weighs 1000 lbs., and the largest even 1500 lbs. ...

The reindeer, or caribou, seldom exceeds 400 lbs. in weight. It is a fleet animal, yet in deep snows it is overtaken by men and dogs. Droves of caribou are sometimes seen sporting upon the mossy bogs of the interior. The flesh is far less palatable than moose venison. The Virginian deer (*cervus Virginianus*), common in New

Brunswick, is not seen in Nova Scotia...[1]

I have remarked elsewhere that the fallow deer (*cervus Virginianus*) was not seen in New Brunswick prior to the year 1810, at which period wolves also appeared — nor has it yet reached Nova Scotia on its march to the east. This beautiful animal has evidently been driven into that province by droves of wolves which are now extending their march in this direction, and far beyond their former limits. In 1845 a wolf was seen in Nova Scotia, on the road between Halifax and Truro. In 1846 two were killed; and more recently others have been captured in the neighborhood of Windsor and Musquodoboit. The Legislature have offered a bounty for their destruction. In my excursions on the interior lakes of New Brunswick, I have seen them in droves, and more than once saluted them with a rifle ball. In the back woods of that province the howlings of those animals during the night are terrific and the sheep of the new settlers are frequently their prey...[2]

The beaver have all been destroyed. Foxes, martins, and other small animals, are hunted and trapped for their furs. Bounties are given for the destruction of the wolf, bear, and lynx, or wild cat. A bear hunt sometimes affords capital sport. Bruin will often walk off with half a dozen balls, and he fights well to the last.

There is still some good goose and duck shooting in the unfrequented rivers during the month of April. Pigeons have not been numerous of late, and the hardy partridge has begun to decline in the older settlements. Woodcock are on the increase. Twenty years ago it was rare to flush a single cock; they now breed in every part of the province, and begin to afford some fine sport. Snipe are numerous on some of the intervales — the borders of the great marshes, and in the alder swamps. Plover of different kinds are very plentiful in some situations during the months of

September, and the great Esquimaux curlew is occasionally seen on the shores...[3]

Formerly the white goose (*anser hyperboreas*) was common in some of our bays during the spring. Not one of them has been seen of late, and they are now rare even in the Gulf of St. Lawrence...[4]

When the French and English first began to frequent Canseau as a fishing situation, walruses, or sea-cows, were numerous, and their teeth, which equal the ivory of the elephant, formed a valuable article of trade. When the ice-fields became closed, these animals would sometimes land and sport in the snow. A century ago they would land upon and cross Prince Edward's Island. Under such circumstances, they were attacked by bands of fishermen with spears, and great havoc was made among them, until they finally disappeared. At North Cape their bones are still found in the forest. The weight of a single walrus would sometimes exceed two tons. They are now said to be on the increase, although I saw only two walruses during my visit to all the island shores...[5]

Porpoises are supposed to pursue the herring, and are seen in droves, lifting and dipping their noses, not only over the whole Atlantic, but close in upon the shores. They are very valuable on account of their oil, but no good plan to capture them has ever been discovered. Porpoise fishing is therefore almost exclusively carried on by the native Indians, who display much patience and skill in this employment. Two of them enter a light bark canoe, and even when the waves are running high, they will paddle out several miles from the shore. The foremast man is ever ready with his gun, and as the nose of the porpoise appears above the water, he fires. The man in the stern then paddles with all his might to reach the animal, for if quite killed, it sinks immediately.

If the shot were successful, the porpoise is carefully

handed into one of the narrow ends of the canoe. As this animal only shows his head above water for an instant the sportsman who shoots him has something to boast of.

The author once accompanied an old Indian in quest of porpoise, and although not a bad one among woodcock and snipe, he missed several shots at these animals. He finally handed the gun to the Micmac, who soon laid his game in the bottom of the canoe, stating very leisurely that 'Porpoise know Englishman, and when he see em he make bow very quick, and say bang away mister, porpoise not all the same one turkey.'

About 1500 gallons of porpoise oil are annually collected at Digby Gut, by a party of Indians from Annapolis and Bear River...[6]

In no part of the world are salmon to be had of finer flavor than in Nova Scotia. In May they begin to visit the rivers and ascend the principal streams to the lakes. But no sooner do they appear than they are threatened with every kind of death. Nets are spread at river mouths to strangle them, or the salmon is brought from his element by the sweep seine. The native Micmac lights his torch, which at midnight renders the fish partially blind, and pursues him along the rippling river, with spear in hand; while the followers of old Isaac Walton throw the deceitful fly upon the surface of the water, to allure him from his lurking place, and make him a prize. Gold River, in the county of Lunenburg, has been celebrated for the sport it affords the angler; but it is inferior to Barrington River, which is now annually visited by gentlemen from the United States, who kill numbers of salmon during their season. The Gaspereau, Maccan, and River Philip offer similar, but inferior temptations to those who are fond of this kind of amusement.

The erection of dams for water mills, across the rivers has proved almost fatal to the river fisheries, and the dust that falls from the saw mills is supposed to be injurious to salmon and alewives. In numerous instances free passages for the fish might have been opened, or the water directed over inclined planes, by which they could ascend to the lakes to deposit their ova: of late this precaution is better observed, but the passages for fish in many streams are obstructed altogether.[7]

Beaver were formerly numerous, but their dams have been destroyed, and it is now stated by the Indians that there is not a single living beaver in Nova Scotia. The extinction of some animals has, however, been followed by the introduction of others. In Nova Scotia the beaver has been succeeded by the wolf, and the skunk (*Mephitid Americana*) has increased in numbers during the last 30 years. But the moose and caribou, and the most valuable fur-bearing animals, seem destined to annihilation. With them will also disappear the native Micmacs, unless some powerful effort is made to bring them within the pale of civilization.[8]

8
Troubles in the Fisheries of Nova Scotia and New Brunswick

Part 1

Of all the resources of British America, there are none more valuable than the fisheries, which, under proper protection and management, would supply the elements of vast export, and consequently of provincial wealth. Their productions may be secured without the outlay of much capital, and as they always command a ready market, there is seldom much risk in fishing enterprise. The utter neglect of the fisheries at many places and the imperfection of the system generally pursued, are proverbial; yet, the inhabitants of the province [of Nova Scotia] view the success of this branch of industry, in its season, with more interest than they do the ripening of agricultural crops, for fish is not only a staple article of commerce, but also one of extensive home consumption. The farmer, therefore, justly considers the annual catch highly important to his interest.

Among all the negotiations that have been carried on by Great Britain with France and the United States, none have been more weak and impolitic than those that relate to the fisheries of British America. The disregard of existing treaties by the people of the republic, who are permitted to fish along the coasts has greatly discouraged the colonists, who are frequently insulted, and driven away from their rightful inheritance.[1]

Part 2

The convention of 1818 allowed the people of the United States to fish along all the coasts and harbours [of Nova Scotia, New Brunswick and Prince Edward Island] within three marine miles of the shore, and to cure fish in such bays and harbours as are not inhabited; but if inhabited, subject to agreement with the proprietors of the soil. It also permits them to enter bays or harbours on the prohibited coasts for shelter, repairing damages, and purchasing wood and obtaining water, and for no other purpose whatever — subject to restrictions, to prevent abuses.

It has been decided by eminent lawyers, that, according to the plain and obvious construction of the convention, the citizens of the United States cannot fish within three marine miles of the headlands of the coasts of Nova Scotia and New Brunswick; and that they have no right to enter the bays, harbours, or creeks, except for shelter, or for the objects before mentioned — and then only on having proved that they left their own ports properly equipped for their voyages: but it has been abundantly proved, by the most unquestionable authority ... that they frequent our shores as freely as they do their own, or as if they had a confirmed right to them ...

Not only do the American fishermen visit our shores, contrary to the terms of the convention of 1818, but they land and purchase bait from the inhabitants. In numerous instances, they set their nets in the coves and harbours of the province, and not unfrequently compel the inhabitants, by force, to submit to their encroachments.

Early in April, schooners, shallops, and other craft, are fitted out in almost every harbour of the eastern states and despatched to the fisheries. They are amply supplied with provisions, salt, empty casks, seines, nets, lines, hooks, jigs, and every article necessary for taking all kinds of fish. Such as are intended for a shore or 'trading voyage' carry a stock of port, flour, molasses, tobacco, gin, and other goods adapted to the wants of the provincial fisherman and his family. As the season advances, the banks and best fishing grounds are covered by these craft, and whole fleets may be seen engaged in drawing up the finny inhabitants of the sea. Very many of these vessels anchor and fish within three miles of the shore. During the evening, they will enter the small bays and inlets, set their nets, and, by early dawn on the following morning, are seen moving off with the fish taken in the dark hours of night. Even farther, when they have been unsuccessful in obtaining bait, they draw and unload the nets of the inhabitants, who, by remonstrating, are almost sure to have their nets afterwards overhauled or destroyed.

Upon the slightest pretext, they take advantage of the humane intentions of the treaty, and enter the harbours, rivers, and creeks to obtain wood and water. On such occasions they frequently set their nets on the shore, and anchor as near the land as safety will admit. Meanwhile their crews are actively employed in fishing.

The vessels, sent out for the twofold object of fishing and trading, boldly enter the harbour, into which they pour their casks of water: they have sprung a mast or boom — one of the crew is sick — or some disaster has happened, whereby they draw forth the sympathies of the inhabitants; but no sooner is the vessel safely moored, than a traffic commences. Green, salted, and half-dried fish are all taken for American goods; which being landed free of any colonial duty, are given to the fishermen at a lower price than those obtained from the established merchant. The work of the smuggler is completed in a few hours; and as he makes his visit at those periods when the fish are most plentiful, he generally departs richly freighted, leaving the flakes and salting-tubs of the shoremen empty. The fishermen of our shores seldom resist these temptations, and they are often deceived by the declaration that their accommodating visitors are true Englishmen. Should a British cruiser appear, or an officer with proper authority take cognizance of the act, some exigency, embraced by the terms of the treaty, is immediately brought to his notice, and duly supported by the solemn declaration of the crew; or if the vessel should be seized according to law, the matter becomes a subject of grave consideration between the two powers, and, forsooth, a war may be threatened by the apprehension of a foreigner taken in the act of carrying on an illicit trade! Thus the resources of the country are nefariously taken away, and the morals of its people corrupted by the introduction of practices which are abhorred by every honest inhabitant.

The merchant who pays the duties on his goods, and advances them to the fishermen of his district under a promise of payment from the fruits of his labour, is defrauded, and the revenue of the province is diminished, by an unlawful traffic. Again, many of our young men are enticed away, and the bounties offered by the Americans to their fishermen are held out as a temptation for them to depart with their chary and cunning visitors.

Such aggressions are not limited to any particular part of the coast, nor to the thinly-populated districts ... On the 5th of June last, an American fisherman was seized while lying at anchor 'inside of the lighthouse, at the entrance of Digby Gut, near the town of Digby, about a quarter of a mile from the shore, his nets lying on the deck still wet, with scales of herring attached to the meshes, and having fresh herring on board his vessel. The excuse sworn to was that rough weather had made a harbour necessary; that the nets were wet from being recently washed, but that the fish were caught while the vessel was beyond three miles of the shore.' [Despatch of Lord Falkland, 1845.]

Throughout the fishing season, their [American] vessels enter the harbours and surround the island of Grand Manan. They are scattered along the shores of the Bay of Fundy, and enter the harbours, bays, and inlets of Nova Scotia, Cape Breton, Prince Edward's [sic] Island, Magdalen Islands, and coast of Labrador, passing into the Gulf of St. Lawrence through the Strait of Canseau ... Their vessels are very numerous in the Gulf, and occupy the best fishing stations on the banks between Prince Edward's [sic] Island and New Brunswick, the Magdalen Islands and the coast of Labrador to the exclusion of British fishermen. In the summer season, they line the north coast of New Brunswick, enter the Bay Chaleurs [sic] with impunity, and frequent the excellent fishing-ground at Miscou Island and those of the Gaspé coast. As the gulf and its bays are almost unprotected by cruisers, they not only fish upon the shores, and carry on an illicit trade, but sometimes drive the inhabitants away by force, take their bait, destroy their nets, and go on shore and plunder the harmless settlers. Such are the operations of the convention of 1818, which was, as an American fisherman compared it to the writer, 'a net set by the British to catch the Yankees; but the meshes are so large, that a fishing craft of a 100 tons burthen might pass through it without touching.'

But the inquiry immediately arises, why are not these aggressions prevented by our men-of-war and cruisers on the North American station? To this is may be replied that none but the smallest class of vessels are suitable for such a service, and of such there are but a few employed on the coast. The whole length of the coasts to be guarded, including their bays and indentations, will exceed 2,000 miles, and therefore protection could scarcely be afforded, except by a large fleet of small vessels adapted for running into narrow estuaries and shallow bays. The movements of a cruiser are well understood by an American fisherman, who, when his vessel is boarded, has sufficient ingenuity, with the aid of the large meshes in the treaty to 'get clear off' ...

In 1839, I had occasion to take passage in a small American mackerel fishing craft, from Eastport to Grand Manan; and, in consequence of the vessel having struck a shoal of mackerel, I was detained 24 hours. The *General Jackson* was filled to the hatches with salt, empty barrels and provisions; along the deck were ranges of empty puncheons and casks, and for each man four mackerel lines, completely fitted, were attached to the inside of the bulwarks. The hook employed is about the size of that used in fishing for salmon, with a conical piece of pewter ingeniously cast in the shank, and kept bright by scouring with the dogfish skin. This is called a jig, and in the water resembles the small sepia, or a kind of shrimp, upon which the mackerel feed. Besides these jigs, there were a number of small iron rods with a hook at the end of each, being attached to a long and light spruce handle. Nets are sometimes used. Several casks were filled with small herring and other fish, in a state of putrefaction — these are used for bait. There is a curious machine called the bait-mill, consisting of a cylinder studded with sharp pieces of iron, and turned in a box, also occupied with knives and wooden pegs. The bait is thrown into the box, the crank turned, and out of a spout comes the ground fish, which is called *poheegan*. While we were dashing along in a pleasant breeze, the crew were employed in preparing bait and cleaning the jigs. Happy in his prospects, one of the fishermen sang 'Jim Crow,' and another chaunted.

> Come, little mackerel, come along,
> Come listen to the Yankee's song;
> See, the day is fine, the cutter's away, —
> Oh, come along and with us play.

After closely observing a large flight of gulls that hung over the water for some time, the old bronze-faced Captain spoke in a mild tone, 'Make no noise. Seth, haul the jib-sheet to windward. Aaron and Washington, small pull main-sheet. Steady, now!'

The schooner now lay driving to leeward, at a gentle rate, when a hogshead of *poheegan* was thrown into the sea, and soon covered the surface of the water with oil and small fragments of fish. The mackerel rose immediately, and formed a close shoal more than three miles in circumference. Scarcely a word was spoken; and, during three hours, all hands displayed the greatest activity in hooking, jigging and drawing in the fish, which sported in millions around the vessel; nor could I remain an idle spectator to the interesting scene. In an instant the mackerel disappeared, and the vessel was put upon her course, having her deck, cabin floor, and every unoccupied space covered with the dead and dying fish, the whole quantity of which was estimated at 25 barrels.

This fishing took place within three miles of the northern head of Grand Manan. Before leaving the hospitable captain, I inquired how he avoided the British cutter, then stationed on the coast? To which he replied, 'Oh, we know how to work them critters to a shavin' don't you see, there are about 300 of us here; every one of us has a little kind of a signal. When any one sees the cutter of your Woman King, up goes the signal; and when the fog is so thick you can cut it into square pieces with a splitting-knife, toot goes the cowhorn (these cowhorns are employed to wet the sails, and are called spouting-horns). You see, this island is 25 miles long; when the cutter comes to one end, we go to t'other; and when she comes to t'other, we go to t'other. Why, friend we bow-poop 'em.'

My voyage terminated, I was landed by the captain, who very politely offered me as many fish as he supposed I had taken during the passage.[2]

Part 3

Some of the American skippers will relate their adventures with all the *sang froid* and good nature imaginable. A very clever old captain told me that he 'once ran into St. Mary's. Tidings of my doings had got out, and on the night of my arrival, a revenue cutter came to anchor right along side of me. What to do I did not know. I could not get away as the wind blow'd a stiffer right into the harbour. All at once I had it. In less than no time I cleared away for action. I sent ashore and borrowed two young calves from one of my old customers, and lifted them on deck. One of them squalled out prodigiously. I stowed all the cordwood I had abaft the foremast; but the best of it all was, I dressed two of my Nantucket boys in women's clothes, topping them off with a pair of bonnets sent in my vessel as a venture. By the first peep of day I set them to washing shirts on deck, and as soon as I seed the crew of the cutter begin to move, which they did not until long after sunrise, I went in my little boat and axed the people of the cruiser if they would give me a bit of old canvass to mend my mainsail, and sure enough they gave me a fairish piece. There we all lay till 12 o'clock, the women washing and drying clothes, and our calves blating like mad for their mothers. The wind came round, the cutter got under weigh, and as she rounded past us, the captain hailed me, and asked if I would sell one of the calves. I told him they were a particular breed and not for sale. 'I think that remark,' he said, 'will apply to your whole crew.' Not a bad joke was it? And after laughing heartily at me, my washerwomen and my calves, my gentleman sheared off. When I seed that his jib was the right way, I made a low bow to him, and after he was clean gone, I sent my calves on shore, turned my washerwomen into boys again, and finished the trade of the *Peggy Ann*.

'In another instance,' said this cute old captain, 'it was stark calm, and as the fog cleared up a little, I saw I was in the very jaws of a ship of war, and I almost gave up all for lost; however, as they were lowering their jolly boat to board me, I skulled off to them, all alone in my little punt, and asked the people in the ship if they knowed what was good for the measles. I could hear him laugh from stem to stern. A big fat man, they called the doctor, told me to keep my patients warm, and to give them hot drinks. It was enough; they took care not to come near the *Peggy Ann* that time.'

By such and similar practices, the merchants who advance to the fishermen goods legally entered, are defrauded, the resources of the country are thrown away, and the morals of the people contaminated ...[3]

It had long been hoped that the remonstrances annually sent from these colonies to the home government, would finally bring protection: but so late as the 19th May, 1845, a despatch to Lord Falkland, then governor of Nova-Scotia, states that 'After mature deliberation her Majesty's government deems it advisable for the interests of both countries, to relax the strict rule of exclusion exercised by Great Britain over the fishing vessels of the United States.' Another despatch of the 17th of Sept., 1845, brought the unwelcome tidings, that the Bay of Fundy 'has been thrown open to the Americans, under certain restrictions.' No policy could be more fatal to the welfare of Nova-

Scotia, than this gradual yielding up of her maritime resources to the neighbouring states.

The abundance and cheapness of bread and other provisions, enable the Americans to fit out their vessels at a lower rate than they can be supplied in any of the ports in the province. The tonnage and other bounties on fish granted by the eastern states have stimulated their inhabitants in this branch of industry, and to their credit it may be added, they are more skilful and indefatigable in taking and curing the finny inhabitants of the sea than the general run of fishermen upon our shores ...[4]

Part 4

The Americans are far more successful in fishing than the inhabitants of the British Provinces, and supply their fish at a lower price than will remunerate our own people. This fact has its origin in a variety of circumstances. Their government affords great encouragement to this branch of industry. A tonnage bounty is given to their fishing vessels, which secures the fisherman against any serious loss in the event of the failure of his voyage. He has also a privilege in the importation of salt, and is protected in his home-market by a duty of five shillings per quintal on dry fish, and from one to two dollars per barrel on pickled fish. The duty imposed on American fish imported into the colonies is much less, and no bounty is offered to their fishermen, whose markets are limited and fluctuating.

In consequence of the great advantages afforded to the citizens of the United States by the treaties, and their ready mode of evading the stipulations of the convention, their whole system of taking and curing fish has been rendered superior to that followed by the people of New Brunswick and Nova Scotia. They have also the advantage of obtaining provisions at a much lower rate, a greater sea-going population, and, from long experience, a better knowledge of the most productive fishing-grounds.

On the coast of the Eastern States, half-a-dozen 'neighbours' will build a fishing-schooner during the winter and send her out manned by a few young men, with an experienced captain and pilot. The proceeds of the voyage are divided among the whole company. If a merchant or any other person send out a vessel, he supplies nets, and the crew find provisions, hooks, lines, &c. On her return, the cargo, or 'catch,' is divided, each of the crew having the share agreed upon (usually five-eighths of the whole). Under these agreements, every man has a direct interest in the success of the enterprise, which stimulates him to industry.

The crews of British fishing-vessels, although equally active, are most frequently hired by the month, and, consequently, they have less interest in the profits of the voyage. Nor is it a rare case that they become disheartened by the threats and insults heaped upon them by their more numerous rivals.

To encourage the fisheries, it has been recommended to admit every article required for them duty-free — a privilege now granted by the Government of Nova Scotia, but one which has been found open to abuses little better than smuggling. A bounty on tonnage, or on every quintal of dry and every barrel of pickled fish, would be returned to the revenue by an increase of trade; and the advancement of agriculture would supply the provisions now imported for the fishing part of the population.

Many of the practices of the Americans, while they add nothing to their own interest, are calculated to destroy the inshore fisheries altogether. It is a very general practice for them to throw all the offal of the fish they take overboard. When such offal is thrown into the sea at a distance of three miles from the land or bays, and in deep water, the consequences are far less injurious than when it is cast overboard near the shore. With a knowledge of this fact, after the offal of several days' fishing has accumulated, the Americans wait a leisure time and throw it into the sea within the range of boat-fishing. The whole mass of 'garbage' is immediately devoured by the fish near the land, and to which it is extremely destructive. (The sharp bones of the spines and heads of fish, when taken by a living fish, penetrate the maw, produce diseases of the liver and death. Fish that feed on offal are sickly and unfit for use.)

By the practice of jigging mackerel, many fish are wounded and finally die, and the living ones always retire from the dead of their own kind. Many undue advantages are also taken of the colonial fishermen, who, from inferiority of numbers, are compelled to submit to threats and insults, and not unfrequently to be driven away from their lawful inheritance. The result has been, that French and American fishing-vessels are rapidly increasing in numbers, while the British fishermen are on the decline; and if the encroachments of those two powers are not speedily and effectually checked, the subjects of Great Britain will be deprived of a most valuable branch of national industry and the government will discover when it is too late, that a most important part of her Colonial resources has been taken away by the aggressions of foreign Powers.

The fisheries of New Brunswick, if duly protected, and pursued with energy, would form one of the principal sources of her wealth and prosperity. The coasts, indented by numerous harbours, bays and rivers, afford every facility for shore and deep-sea fishing; and although the practices of the Americans have annually reduced the numbers of the finny tribes, they are still sufficiently numerous to render the employment, under proper management, profitable. But, from causes already adverted to, the demand for timber, and a scanty population, the fisheries are not pursued with energy, and the fishermen lack the stimulous of the bounties given to the Americans, with whom they are unable to maintain a competition.

The whole number of fishing-vessels belonging to the ports and harbours of the Bay-of-Fundy side of the province, in 1840, was only 65. Their burthens were from 10 to 30 tons each. The present number, including 20 belonging to Grand Manan, will not exceed 70, exclusive of shore fishing-boats. That island alone, with a proper population, could employ advantageously 100 and the whole coast 600. The number of fishing vessels belonging to the United States, and fishing in the same waters, is as ten to one. The fishermen of the province, with few exceptions, are far less persevering and industrious than the Americans, or even the people of Nova Scotia.[5]

9
Farming in Nova Scotia and New Brunswick

Few people in the world live better than the farmers of New Brunswick. By their industry, they raise an abundance of agricultural produce; and they have been censured for their extravagance in consuming the food that would bring a high price in the market; and by the sale of which their gains would be increased: but they reply that none have a better right to enjoy the fruits of the earth than those who toil for them. Three bountiful meals are provided every day, and these are often followed by a hearty supper. Their tables are generally well supplied with beef, pork, mutton, and fowls, with pickles, and a variety of vegetables. Salmon and other kinds of fish are also provided; with these there are tea, coffee, cakes, pies, gingerbread, and almost every luxury it is possible for the country to afford. They are not very social in their habits, and their manners are unpolished; but if a friend or stranger put himself in the way of their hospitality, he will find good fare and a hearty welcome: indeed, any person of respectable address and appearance, who can tell a good story, sing a good song, and play the fiddle, may travel through Nova Scotia and New Brunswick free of expense; nor will he always lack the means of a comfortable conveyance from place to place, or hose or mittens when the weather is cold; and the farther he keeps from the towns, the more successful will he be in his economical tour...[1]

Chemistry and geology are ... the handmaids of agriculture. The ash of one crop is very different in kind and quantity from that of another. The ash of wheat contains from 18 to 20 per cent of potash, but the potash in the ash of oats is only about six per cent. According to these facts, wheat will exhaust the soil of its lime and potash much more rapidly than oats; and this demonstration, with many others of a similar kind, is corroborated by practice. Soils contain the above minerals in greater or less proportions; plants also receive them from the earth in different quantities. The experience of the farmer has taught him to apply certain plants to certain fields; but chemistry is the proper guide in fixing their habitations: hence also we deduce a natural reason for the rotation of crops. Many are surprised that felled forests of hard wood [sic] are succeeded by spruce and fir; but nature evidently directs this rotation, and only allows such trees to spring up as the surface of the earth is able to bring to perfection. A backwoodsman once told me that he raised wheat and potatoes upon a piece of ground until they would grow no longer; he then 'pitched it out for a rabbit pasture and cleared a new bit.' Such rabbit pastures are seen in every part of Nova Scotia, and the change of crops displayed by the forests is disregarded in the practice of husbandry.

The leaves that fall from the woods in autumn give the earth an annual top-dressing. A part of the inorganic matter taken up by the trees is thus returned to the soil which is constantly renovated. This fact gives the farmer a useful hint. There are certain tracts to which gypsum might be most beneficially applied — others would be benefited by the carbonate of lime. Some soils require the introduction of vegetable matter, some lack the saline ingredients. These, and many other important particulars, are not understood. Immense quantities of valuable manures are lost for the want of that general knowledge of chemistry and geology so requisite to preserve them, and to guard against their improper application. The causes and effects well known to science are disregarded, and much labor is lost in experiments performed in ignorance of the most obvious laws of nature.

Practical chemistry, geology, and the first principles of agriculture, are untaught in any of our colleges or schools. The physical sciences are not cherished by the legislature, and the few individuals who have devoted their time and energies to the most useful inquiries have labored under every discouragement.

Among the resources of the country, manures, or the means afforded for renovating the soil, are of importance, especially in districts where agriculture must form the chief branch of industry. In few countries are manures more abundant and varied than in Nova Scotia. There is scarcely a tract to be found which may not be improved by its own lime, peat, or sea-weeds...

Animal substances act powerfully upon the soil, especially during their decomposition. In Hungary, and other parts of the world, dead flies are collected and employed for manure. In Nova Scotia the undigested animal manures consist almost altogether of dead fish and fish offal, and these are often allowed to be wasted in heaps rather than applied to the soil. Herring are sometimes taken upon the coast in such quantities that the fishermen are unable to preserve the whole catch, and with smelts and capelin, they are taken for the sole purpose of supplying manure. The taking of fish for this object is very reprehensible, and should be prevented by law, for it tends to destroy the young fry and the bait that allures the deep sea fish to the shores. Herrings, smelts, or capelin, are taken upon the shores of New Brunswick and the coast of Gaspé in great quantities and carted into the fields. This unprepared manure produces a few good crops, but by its constant application the soil loses many of its natural properties, and finally becomes very meagre...

The organic and earthy matter of bones have long been known to be powerful stimulants to the soil; and as a manure, bonedust is perhaps equal to guano. There is not a bone-mill in the province except at Horton. Bones are shipped from St. John, New Brunswick, to Great Britain, where they are ground for manure...[2]

Dry vegetable matter, straw, weeds, or hay, can be made to ferment, and consequently to act upon the soil. The decay of saw dust [sic] is slow, yet it enriches the earth where it is applied. The numerous saw mills in the province produce great quantities of saw dust [sic], which is washed away, and lost.

The powder of charcoal will absorb noxious vapours and oxygen from the air, and also take up the impurities of water. It is therefore of much value for mixing with liquid manure and night soil. In a country abounding in wood, charcoal may be advantageously employed in agriculture. The stagnant water seen near farm houses might be converted into rich fertilising [sic] matter, and the air around them rendered pure and healthy by the employment of charcoal...[3]

In a country where the winters are long and severe, the grasses are of much importance to the farmer. The natural grasses are very numerous, and occupy all the unimproved pastures and intervales. Many of the low grounds and wild meadows supply fodder of wild grass; they are, therefore, very advantageous to the new settler. The marshes also have their grasses; and in summer, cattle and sheep find their livings in the green herbage of the forests. Of the artificial grasses, clover, timothy, and brown top, are considered the most valuable. On some of the dyked marshes the former will yield from two to four tons of hay per acre. The after, or fall feed on such lands, is peculiarly valuable for fattening cattle. On the uplands, the red clover is sometimes uprooted by the severity and heaving of the frost.

A rotation, embracing clover, and the white and green crops, is seldom practised. Thousands of acres of pasture lands might have their value enhanced by a single ploughing, which, with the cost of fencing, would be compensated by a crop of oats, sown with clover and timothy in the spring. The highland pastures of wild grass are excellent, and in autumn they frequently return the stock of cattle and sheep that feed upon them, in high condition...[4]

That the agricultural societies in the different counties of the province have been in some degree beneficial to the farming interests, and especially to the improvement of the breeds of domestic animals, there can be no doubt; but not more than one farmer in ten has been induced to enroll himself; and when party and political feelings are not the cause of their remaining aloof, they generally believe that the kind of information best calculated to aid them is not to be obtained through such channels. True it is that the spirit of inquiry slumbers in their bosoms, and they choose to tread the old beaten track of their fathers, rather than avail themselves of modern discoveries.

The agriculture of Nova Scotia has been depressed by the lack of markets and facilities for transportation. There are no public works, nor manufactories [sic] to promote the local consumption of agricultural produce. The requirements of the City of Halifax, and the small inland and sea-board towns, are altogether inadequate to the supply afforded by the agricultural surface, which might be rendered capable of yielding exports of bread stuffs. There are no railways, nor canals, and the cost of transporting the produce of the farmer paralyses [sic] his energies. These causes alone have induced many of our hardy yeomanry to emigrate to the United States, where they find brisk markets, and a cheap transport for the produce of their labour.[5]

10
Minerals, Fossils and Oil

Of all the resources of Nova Scotia, there are, perhaps, none of greater value and importance to the permanent welfare of the country than those of the mineral kingdom. But from their nature and situation, they are the most difficult to discover and bring into operation. Timber may be felled — fish may be taken, cured, shipped, and vegetables may be cultivated, by solitary individual industry, and almost without the aid of science, or capital: hence it is that the most common natural objects of a new country afford its first exports. Not so with mineral productions; — their discovery — their removal from their native situations, and their application to useful purposes, require science, skill, and the outlay of great capital.

From the first discovery of Acadia, or Nova Scotia, up to the present time, the mineral productions of the province have been considered of great value. During the first voyage of DeMonts into the Bay of Fundy, in 1604, he found native copper at Cape d'Or, and amethysts, with other gems, at Parrsboro and Cape Blomidon. One of these gems was placed in the crown of the king of France...[1]

Mining scarcely forms a part of provincial labour, for it is conducted solely by the General Mining Association...[2]

Of all the minerals found in the earth, coal and iron administer more to the necessities and comfort of mankind than any other. Coal not only supplies domestic fuel, but it sustains the furnace and the forge. In the production of steam it propels machinery, and by it many of the multifarious operations of manufacture are performed almost without the aid of animal labor. The application of steam to navigation, and the construction of railways, have brought about a new era in the affairs of the world, and thereby the importance of coal is daily on the increase...

The coal of Nova Scotia is sufficient to supply the whole steam navy of Britain for many centuries to come, and also to meet amply the demands of the North American Colonies. That resources of so much value should be left free and open to public competition — to public enterprise — is necessary to their development and beneficial application; yet certain it is that the inhabitants of the province have never evinced much mining energy...[3]

When the quality of the Cumberland coal is considered, and its geographical situation properly viewed, it is remarkable that those who claim an extraordinary right to the Mines and Minerals of Nova Scotia [the GMA], should neglect a situation offering so many advantages. Almost all the coal raised from the mines at Pictou and Sydney, is transported to the United States, where its demand is steadily increasing. But vessels loading at those places, must almost circumnavigate the province, before they can obtain their cargoes. On the other hand, if they were supplied at Cumberland, the length of the voyage would be greatly diminished, and consequently coal would command a higher price at the pit's mouth. It is however very probable that the Mining Association, having expended large sums of money at the coal mines of the eastern parts of the province, would rather discourage the coal trade from Cumberland than abandon their former labours. And while competition is prevented, and the inhabitants of Nova Scotia are only permitted to gaze upon the treasures of their country without being permitted even to dig a bushel of coals from beneath the soil so long in all probability will the mines of Cumberland remain closed, although steamboats propelled by fuel brought from England shall almost daily pass over their strata...[4]

The Cumberland coal field may justly be called a vast fossil valley, where plants from the lowly iris, up to the majestic palm, have been buried by some great and sudden change on the surface of our planet. The area included within the limits of this singular event is by no means narrow or confined to the petrifaction of a few lignites: it reaches at least 15 miles along the shore, and more than 20 into the interior of the country. The banks of rivers and creeks, the sides of ravines and cliffs, have been examined, and the same fossils are everywhere exposed over several miles on the surface: and even among the common rocks of the field the remains and impressions of antediluvian plants are yearly overturned by the movements of the plough and hoe. These facts should be remembered, as they plainly show that no common causes could have produced effects so wide in their operations, and powerful in their results.

Between the Bank Quarry and the coal veins, there are sections of two large fossil trees, standing perpendicular by the side of the cliff, and penetrating the strata in their way upwards: but as the precipice is constantly yielding before the action of the elements, its strata have fallen, and in their descent carried downwards large portions of these trees, which may now be seen among the numerous relics of the shore...

A few miles southward of the 'King's Vein,' we discovered an immense fossil *Lepiododendron Aculatum*; the violence of the sea had removed the adjacent shale, and sandstone, and the majestic plant remains by the side of a vertical cliff. This tree stands perpendicular, passing through and crossing the strata, according to the angle of their dip. Its roots are seen branching out, and penetrating the rock beneath. At the base it measures two feet eleven inches, and 40 feet of its trunk were exposed at the time of our last visit to the spot. Sections of a still larger growth may be seen along this unfrequented shore, and pieces of smaller dimensions may be observed from 50 to 100 feet up the embankment.

Frequently the bark of these trees is converted into coal, constituting the true lignite; in other instances the bark, with the tree itself, is changed into compact sandstone. Great care should be taken in removing pieces of the former as sometimes a whole tree, having its cortical portion carbonized, will sip through the bark and come headlong to the beach. In this way we were in danger of being killed from the unexpected launch of a huge fossil.

Since a recent visit to the Joggins our agent in fossil affairs, a sturdy miner, has informed us that a portion of the cliff has lately fallen, and exposed another tree of great size. But a few days have elapsed since we found a gigantic plant imbedded in the sandstone at low water mark, opposite the Bank Quarry: it had been exposed by blasting the rock for grindstones, and the miners suffered some loss and disappointment, in consequence of its passage through a profitable layer of stone. At this place a *cactus*, beautifully figured on the surface, and measuring 15 feet in length, had been broken by the workmen, and rolled off the reef. Such are some of the ponderous fossils of this valley, to which months might be devoted in collecting and describing the remains of a former world...[5]

From the facts already described, perhaps the following theorem may be derived; — that the Cumberland coal field was at some very remote period covered with a very luxuriant tropical herbage; during the growth of enormous plants upon its bosom, it was by some Geological catastrophe buried beneath the waters of the ocean...[6]

❋ ❋ ❋ ❋ ❋ ❋ ❋ ❋ ❋ ❋

The asphaltum of New Brunswick, now called Albert coal, is one of the richest materials ever discovered for the manufacture of oils. Seventy per cent of the first distillate, after purification may be brought up to a specific gravity of 0.820, and burned in the ordinary coal-oil lamp. The material contains nitrogen, and therefore yields ammonia. It melts in the retort, and the volatile parts escape at a lower heat than those of coal. This may account in some degree for its greater yield of oils and their freedom from impurities. From it naphthaline is seldom produced; and although paraffine is found among its products, creosote and other compounds of its class exist but in small quantities, while the illuminating oils are abundant. The oils themselves belong to a series which contains less carbon than ordinary coal oils. They burn freely, and give a clear, white light...[7]

When oils were first distilled from coals, few attempts were made to free them from their offensive odors, or remove their coloring matters. The only mode practised was fractional distillation, which is altogether quite ineffectual for that purpose. Although the oil made by the Earl of Dundonald in 1781 was burned in lamps, it does not appear that any process of purification was practised at that time. The earliest mode of purifying petroleum was simply to distill it with water, and this is more beneficial than some of the modes practised in the present day, by which the characters of the oils are changed and their illuminating powers deteriorated...[8]

The oils from different coals require different treatment. The oils of Albert coal (asphaltum), Boghead and Breckenridge coal are easily purified; while the oils from ordinary American, English, and Scotch cannels, require more skill, the cost more to bring them up to a fair standard among the hydro-carbons sold in the market.

The author has made more than 2000 experiments in reference to the manufacture and purification of oils distilled from coal, petroleum, and other materials ... Improvements are constantly advancing, and some time may elapse before their manufacture is brought to perfection and the distilled hydrocarbon oils attain that commercial and economic value they are destined to reach...[9]

In the early operations of manufactories [sic], and more especially when their chemists have not had the advantages of experience, residual and resultant products are frequently overlooked, attention being directed altogether to the staple articles called for in the market. This remark is peculiarly applicable to coal and petroleum oil manufactories [sic]. In the United States, at the present period, vast quantities of pitch, heavy oils, and other valuable refuse products are permitted to flow into creeks and rivers, being considered worthless. It is to those products the manufacturer should direct his care and attention. To their careful use he will have to look ere long for no small part of his gains. Competition will ultimately reduce the actual profits to be made upon the oils themselves within narrow limits, and success will only be awarded to economy and superior manufacturing skills.[10]

11
Industry in Nova Scotia

Nova Scotia abounds in the elements of manufactures, and with her inexhaustible supplies of coal and iron, her facilities for propelling machinery, and an agricultural surface capable of supporting a dense population, her advantages excel those of any of the eastern states of the American union; but her industry must be cherished and protected, or her active sons will continue to leave her shores and seek foreign employment. It is vain to suppose that a free trade system will be beneficial to a new and struggling colony which has nothing to export but raw materials; it is rather calculated to enrich an old commonwealth, whose people by their skill and labour make such raw materials valuable, and then return them for consumption. The result of the system alluded to has been that the suppliers of the raw material at last become hewers of wood and drawers of water to the manufacturers.

At the present moment the Americans take from us

gypsum, grindstones, and other unmanufactured articles, (agricultural produce excepted,) at a low, or almost nominal duty. They have recently reduced the duty on coals one half, because they require the article of our hands to maintain their manufactories [sic] and steam navigation; but if our plaster be calcined and ground — if we even polish our grindstones, which are now cut over by themselves — we are told distinctly that we shall not manufacture for them, and upon every article thus manufactured by us they levy a duty of from 20 to 30 per cent. Nova Scotia, on the other hand, receives the manufactures of the United States at a low rate of duty. The consequences of such a state of commerce have been rendered manifest by the almost utter destruction of our infant manufactories [sic], and the emigration of our tradesmen. Certainly this is not a free trade, nor a reciprocal trade, of late so much boasted of...[1]

The manufactories of Nova Scotia have been chiefly confined to the simple operations of sawing wood into deals, boards, laths, and shingles, which are shipped to Great Britain and the West Indies. Excellent flour mills have been erected in the neighbourhood of Halifax, Liverpool, and Annapolis, for the manufacture of flour from foreign grain; but their operations are unsteady, being always affected by the fluctuations of commerce, and a non-protective tarriff. The numerous grist mills in the country villages are only employed in grinding the grain raised in the province for the domestic supply, and they are usually very imperfect in their construction. In all the western counties there is a lack of oatmills. Many of those that were erected under the bounty of the legislature were suffered to fall into decay as soon as that bounty was received by their proprietors.

The distilleries and tanneries in the vicinity of the capital are not in brisk operation. Some of the former have been closed in consequence of the high duty levied upon their productions, and the admission of foreign spirits. Tanning is often conducted by the farmers themselves; yet, there are small tanneries in every county, and the trade is frequently combined with shoemaking.

Excellent castings are made at the iron foundries of Halifax, but all the iron employed is imported from Great Britain, notwithstanding the province abounds in the best varieties of the ore.

A pail manufactory was recently established on the Truro road, 12 miles from the city of Halifax, and large supplies of its productions have been sent to different parts of the province. Small potteries have been successful. The United States market is supplied with grindstones from the county of Cumberland. The grindstones of the bank quarry at Minudie, owned by Amos Seaman, Esq., are superior to any others ever discovered on the continent of America. Other manufactories [sic] have been attempted in the province, but they have so far been unsuccessful — a circumstance that may be ascribed to the high price of labor, and the want of sufficient capital to bring each operation to perfection...[2]

On an average, 80,000 chaldrons of coal and 50,000 cords of wood are shipped annually from Nova Scotia to the United States, which return large quantities of manufactured iron. Implements of husbandry, stoves, culinary utensils, edge tools, and even the axes employed in felling the forest, are imported from the Americans. The manufacture of iron in Nova Scotia would scarcely affect the iron trade with Britain; but it would operate in diminishing the imports from a foreign power, which levies high taxes upon agricultural produce, and all goods manufactured in these provinces.

It is to be regretted that the Annapolis Iron Company had not employed men of science and skill; for beyond the selfish motives of those who had the control of the works, it is evident that they were not practically acquainted with the art of smelting iron. With a large furnace, and a powerful blast carried by water, they were only able to obtain, with the best charcoal, 13 tons of cast iron per week — not equal to one-third of the produce of the English and Scotch furnaces, which work ores of nearly the same per centage...[3]

There are at present seven iron foundries in the two provinces above mentioned [Nova Scotia and New Brunswick]. Those foundries are supplied with iron from Great Britain. Now, if the proprietors of those foundries can import their iron, and manufacture it into castings, under the high rate of wages of the country, and make sound profits, it is obvious that the ore of those provinces may be smelted with profit also, especially as the important article of fuel, either wood or coal, may be obtained cheaper here than in any other inhabited country. From the scarcity and high price of iron in Nova Scotia, and indeed in all the British American provinces, its use is limited in agriculture and all kinds of machinery; and if any of our ships have been imperfectly fastened, the fact has resulted from the high price of iron, and not from design or negligence. All the iron employed by the Mining Association, for railroads and other purposes, is imported from Great Britain, and having been transported 3,000 miles, it is finally thrown into castings at the very site where thick beds of Nova Scotia ore are seen protruding from the earth, and where a single stratum of coal, 36 feet in thickness, is ready to supply the fuel requisite for smelting and manufacture...[4]

In 1826 a company was formed at Halifax to open a canal from Dartmouth to Shubenacadie, through the lakes, and across their barriers. As nature had almost completed the communication, this project was undertaken with much spirit, and high expectations of success; but, like many other attempts at public improvement in these colonies, this enterprise, which cost £80,000, terminated in a failure, after the work itself had been nearly completed; and now that railways have succeeded canals, it is not probable that it will ever be resumed. The great error appears to have been the lavish expenditure of money upon works of masonry that might have been chiefly constructed upon the American plan — the employment of brushwood, timber and earth. Some engineering difficulties also appeared that were not understood at the onset. A general panic seized the shareholders of the company, and they withdrew from an undertaking that was perfectly practicable. The fine masonry of the canal is now falling down; and a number of persons who had embarked their capital in the enterprise have been nearly ruined ...

The failure of this undertaking, and the Annapolis iron works, have had a powerful tendency to discourage all enterprise of the kind; and an opinion prevails up to the present moment that it is too early to introduce any extensive public works, except common roads, into the province ...

No less than 20 rivers fall into the [Minas] basin ... In their descent, they offer abundant power for the manufacture of the productions of the country...[5]

It has been frequently imputed to the inhabitants of Nova Scotia that they have less perseverance, enterprise, and industry, than the Americans of the republic. Admitting the correctness of this opinion, it should be remembered that very many of her early settlers were several years engaged in the defence of their country: and many of them, on account of their loyalty, abandoned the cultivated fields of their forefathers in the now United States to cut down the forest a second time in order to win a living. They were an exiled people, who had to encounter all the difficulties of colonization in a climate unmodified by the spreading out of cleared fields and the redemption of extensive marshes ...[6]

It may be admitted that the productions of the different tradesmen are well and substantially made; yet they ordinarily lack lightness, finish, and the ingenuity peculiar to articles manufactured in the United States, to which a great many of our best mechanics emigrate annually. Numbers of wagons, and other kinds of carriages, farming utensils, wooden clocks and household furniture of every description, are annually imported from Boston, and other American ports. They at once recommend themselves, and are purchased at high prices, in preference to any made by the mechanics of the country. The consequences of this state of things are obvious, and will soon materially diminish the use of British manufactured goods. No general effort has been made to encourage our own manufactures in preference to those of a foreign power, which, by heavy duties, exclude almost all our productions, except coal, gypsum, and other articles that they cannot obtain from any other quarter.

It is a reproach to these British provinces that, besides immense supplies of bread stuffs, they import from the United States numerous articles that might be as cheaply manufactured within their own boundaries ...[7]

The time has arrived when it has become necessary to direct the public attention to the industrial resources of Nova Scotia; for it is to such objects the inhabitants must apply their labour, to meet the great outlay required for these costly undertakings ...[8] but unless the government enter deeply into such improvements, justly viewing them as safeguards to these colonies, the struggles of the inhabitants themselves will be unavailing, and Nova Scotia will languish through centuries to come.[9]

12
The Gold Fields of Nova Scotia[1]

Tangier

In 1860 gold was accidentally discovered several miles northward of Tangier River, an inconsiderable stream flowing into the Atlantic, 50 miles eastward of Halifax. In March, 1861, gold was found in a small brook a mile eastward of the mouth of the river. When these facts became known, persons who had been engaged in mining in California and Australia, with others, immediately rushed to the spot and commenced the extraction of the gold from the rocks and the sands of the rivulets. It is to those miners the province is chiefly indebted for the knowledge that has been won of the gold in other parts of the country. The rocks in which the gold appears at Tangier are quartzite, and metamorphic clay slate and graywacke ... A considerable quantity of gold has been taken from the quartz boulders found in the drift of this region. The strata contain numerous veins of quartz which, when they run parallel to the beds, are called "leads." These veins are from one quarter of an inch to 15 inches in thickness; besides them there are cross veins containing gold. Most frequently these quartz veins are enveloped in a thin coating of clay slate. In breaking the quartz the gold is found associated with iron, copper and arsenical sulphides, thin scale, and mica and talc, or chlorite ... The gold occurs occasionally in the clay slate, and it has been washed from the soil. It is imbedded in the quartz in minute particles, or in pieces weighing from a few grains to some ounces. One nugget was sold for $300 ...

A lake has been drained by a company formed for that purpose, and it is probable that its bottom will afford rich washings. During the writer's visit to the spot, a mass of slate weighing 70 lbs. was raised from the side of the lake, and was filled with innumerable grains of gold, weighing from one half a grain to two grains.

A street of shanties runs through the "diggings": And there are small inns for the accommodation of strangers. Upwards of 200 claims, each 20 by 50 feet, have been granted by the government, besides several of three-quarters of an acre each. Upon these claims 600 men were employed during the summer of 1861; but on the approach of winter the greatest number of the diggers returned to their homes to prepare for the coming season ...

At Strawberry Hill three tunnels have been run into the side of the eminence, and mining has commenced systematically ... The amount of gold already mined at this place cannot be accurately ascertained, although it is known to be very considerable. In general the miners have made wages upon their small lots, and a number of the more fortunate ones have won from $300-$500 each, over and above their expenses, during the past summer [the summer of 1861] ...

Lunenburg — "The Ovens"

"The Ovens," now celebrated for gold, are four miles by water, and 13 miles by land from the latter town [Lunenburg], and on a point of land jutting out into Lunenburg Bay, a most spacious and safe harbor. "The Ovens" are inconsiderable excavations worn out of the rocks by the sea, and they have given a name to a peninsula about a mile in length and half a mile in breadth...

The quartz veins of the little headland and the slates contain gold, which is here also accompanied by the sulphurets of iron, mispickel, mica and the oxides of iron. From the more rapid disintegration of the slates, a small cove about 250 yards wide has been worn out by the sea, and into which the sand of the shore and debris of the slate with gold have been collected.

By being washed this sand has already supplied much of the almost pure metal, and it still continues to be productive. During the past summer a few lots of 25 by 50 feet each, were sold by Mr. Cunard for $4,800, under a reservation of one-quarter, and after, as it is believed, a large amount of gold had been removed.

It was also stated that ten men obtained 18 ounces of gold by hand-washing, in a single day. All the sands of the adjacent shore contain gold, and during the past summer washing has been carried on with great industry. The quartz veins of the higher grounds are also productive. The quartz was piled in heaps, awaiting the introduction of crushers and other means of extracting the gold from its gangue. Several nuggets have been discovered. The largest of these shown to the writer was found in a crevice in the slate at the south shore. It was attached, on one side, to a piece of red dish-colored quartz, and weighed one and a half ounces...

The shore lots leased by the government were only 20

Goldwashing near Lunenburt (at The "Ovens"). *The Illustrated London News*, October 5, 1861. Courtesy Special Collections, Killam Memorial Library, Dalhousie University.

by 50 feet in area; but a number of lots laid out on the uplands embrace each three-quarters of an acre. The whole amount of gold obtained from the "Ovens" of Lunenburg from June to December, 1861, is estimated at $120,000, which has been separated without the assistance of any kind of machinery. All the lands on the little peninsula have been taken up, or applied for, and upwards of 30 companies have been formed to work the gold in this part of the country.

It has been supposed that the golden sand was washed up by the sea. Of this the writer found but few evidences. All the rocks beneath the lowest level of the tide are thickly covered by sea weeds. The sands and gold which collect upon the border of the water line have, evidently, been derived from the crumbling down and decomposition of the rocks of the shores, and the softer slates have yielded more rapidly to disintegration than the quartzites. So far, therefore, their water front has produced the greatest quantity of the metal. It will be found hereafter that each succeeding winter, with its severe frosts, and spring, with its alternate freezings and thawings, will produce an annual crop of the precious metal from the soil...

About 800 men were employed at the above place during the month of September last. In October that number was reduced to 200 on account of the approach of winter. About 40 small buildings have been erected at the "diggings," with an adequate number of hotels...

Parallel Lines
The gold of Nova Scotia appears chiefly to exist in certain parallel lines, which probably extend in some instances almost the entire length of the province, or to the distance of 200 miles in the direction of the strata...

It is a peculiar feature of the gold regions of Nova Scotia that regarding recent discoveries the rocks containing the gold in the highest per centage [sic] are near the Atlantic coast, and intersect a number of the smaller rivers and harbors, whereby facilities are afforded to supply the requirements of mining. It is not at all probable that the richest gold deposits in Nova Scotia have yet been discovered; but there is enough known to satisfy the most skeptical that the province contains an ample amount of the precious metal to warrant the most extensive operations and the construction of machinery for its mining and purification.

Assay of Gold
An assay of a sample of gold from Tangier gave the following result from 100 parts:

Gold..........	96.50
Silver.........	2.00
Copper	0.08
Lead..........	0.06
Iron	0.05
	98.69

The gold from the "Ovens," Lunenburg:

Gold..........	93.06
Silver.........	6.60
Cooper	0.09
Iron, a trace	
	99.65 [sic]

Nuggets of gold, weighing several ounces each, have been found by the miners at Wine Harbour, Tangier, Laidlow's and the "Ovens." The largest of these have been purchased by the Nova Scotia Government, to be displayed at the approaching Industrial Exhibition of London.

It was gratifying to observe the good order that prevailed everywhere among the miners, and the ability displayed by the surveyors who superintend each mining district. Property of all kinds is perfectly safe, and the provincial authorities are ready to encourage those who desire to embark in the development of the gold of Nova Scotia.

January, 1862

Notes

Introduction

1. *The Scientific American,* October 12, 1850.

2. See Daniel Yergin's bestseller, *The Prize* (New York: Simon and Schuster, 1991), 23.

3. See chapter 8 in this book.

4. G.W. Gesner, "Dr. Abraham Gesner — A Biographical Sketch," *Bulletin of the Natural History of New Brunswick,* No. XIV (Saint John, N.B.: 1896), 6-7.

5. Ibid., 10-11. George Gesner remarked in "Dr. Abraham Gesner — A Biographical Sketch" that his father "was very fond of music and quite a capable performer on the flute and violin in the family circle." Moses Perley refers to Gesner playing the bugle in "Sketch VI" of his "Sporting Sketches," published in London in the *Sporting Review* in the early 1840s — and reprinted in *Camp of the Owls,* edited and illustrated by Peter Mitcham (Hantsport: Lancelot Press, 1990), 68-69. The many significant links between Perley and Gesner are

traced in my books, *Three Remarkable Maritimers* (1985) and *Abraham Gesner, Prophet of the Wilderness* (1995).

6. G.W. Gesner, "Dr. Abraham Gesner — A Biographical Sketch," 7.

7. Moses Perley, "The Indian Regatta," *Camp of the Owls*, 68.

8. George Gesner says of his father in "Dr. Abraham Gesner — A Biographical Sketch." He was always abstemious and temperate in his habits of life, but liked and would smoke a good cigar," 10. Given the customs of the time, being abstemious was unusual. In the *Journal and Proceedings of the House of Assembly of Nova Scotia* (1849), in the "Aggregate List of Articles purchased for the use of the Halifax Asylum for the Poor during the year 1848," beer, 6398 gallons of it, totaling £51 3s 6d, out of an expenditure of just over £3000 for all purchases and salaries — was third on the list of *necessities*. In addition, "Wine for the Sick" — 87 gallons, costing £22 5s 11d, was purchased. (See Appendix 40.) The beer and wine together were considerably more than Dr. Almon's £50 salary for attending to the sick in this poorhouse.

9. *Abraham Gesner, Prophet of the Wilderness*, (Hantsport, Nova Scotia: Lancelot Press, 1995).

Chapter 1

1. Abraham Gesner, *The industrial resources of Nova Scotia; comprehending the physical geography, topography, geology, agriculture, fisheries, mines, forests, wild lands, lumbering, manufacturies, navigation, commerce,*

emigration, improvements, industry, contemplated railways, natural history and resources of the province (Halifax: A. & W. MacKinlay, 1849), 27.

2. Ibid., 34 & 35.

3. Ibid., 27 & 28.

4. Ibid., 28-30.

5. Abraham Gesner, *Remarks on the Geology and Mineralogy of Nova Scotia* (Halifax: Gossip and Coade, 1836), 242 & 243.

6. Ibid., 244 & 245.

7. Ibid., 116 & 117.

8. Ibid., 248 & 249.

9. Ibid., 245 & 246.

10. *The Industrial Resources of Nova Scotia*, 300-310.

11. Ibid., 41.

12. Ibid., 42.

13. Ibid., 41.

14. Ibid., 38 & 39.

15. Ibid., 44-46.

16. Ibid., 133. This quotation continues: "Frozen fish are also shipped from other ports. Salmon are purchased at Medway, LaHave, and Gold rivers, and being packed in ice are sent to the United States, where they command high prices."

17. Ibid., 44-46.

18. Ibid., 46 & 47.

19. Ibid., 47.

20. Ibid., 47 & 48.

21. Ibid., 53 & 54.

22. *Remarks...*, 172 & 173.

Chapter 2

1. Abraham Gesner, *New Brunswick with notes for emigrants* (Saint John: Chubb, 1847), 67-69. It is no accident that Chapters 1 and 2 both begin with Gesner's observations on the Fundy coast. In both provinces the land lying in proximity to the Bay of Fundy caught and held his interest more than any other.

2. Ibid., 70.

3. Ibid., 71.

4. Ibid., 81.

5. Ibid., 174.

6. Ibid., 84.

7. Ibid., 85.

8. Ibid., 92.

9. Ibid., 93.

10. Ibid., 95.

11. Ibid., 90-91.

12. Ibid., 134-135.

Chapter 3

1. All the material in this chapter has been taken from Gesner's "Report of the Geological Survey of Prince Edward Island," Appendix D, *Journal and Proceedings of the House of Assembly of Prince Edward Island*, 1847. Since the pages are unnumbered, specific page references cannot be made.

Chapter 4

1. This chapter is an excerpt from the most detailed of Gesner's reports on the Micmacs when he was a commissioner of Indian affairs. It was originally published in the *Journal and Proceedings of the Nova Scotia House of Assembly*, Appendix 24, 1847.

Chapter 5

1. Abraham Gesner, *New Brunswick with notes for emigrants* (Saint John: Chubb, 1847), 108. The spelling of 'Melicite' which Gesner uses is variable, 'Malecite' perhaps being the most generally accepted form today.

2. Ibid., 110.

3. Ibid., 111.

4. Ibid., 112.

5. Ibid., 114.

6. Ibid., 117.

Chapter 6

1. Abraham Gesner, *The Industrial Resources of Nova Scotia...* (Halifax: A. & W. MacKinlay, 1849), 73-74. The wealth derived from exploitation of forest resources in New Brunswick had for many years previous to Gesner's observations superseded that of Nova Scotia. Haliburton remarked in *An Historical and Statistical Account of Nova Scotia, in two volumes* (Halifax: Joseph Howe, 1829, Vol. 2), 107 that "The value of Imports of New Brunswick exceeds those of Nova Scotia this year, in consequence of the great advance in the timber trade, 75, 978; and the Exports exceed those of Nova Scotia, 54, 665.

2. Abraham Gesner, *New Brunswick with notes for emigrants* (Saint John: Chubb, 1847), 238.

3. Ibid., 168.

4. Ibid., 187.

5. Ibid., 190.

6. *The Industrial Resources of Nova Scotia*, 100-101.

7. Ibid., 96-97. See pages 77 ff. for a catalogue of the indigenous trees of Nova Scotia.

8. Ibid., 214-215.

Chapter 7

1. Abraham Gesner, *The Industrial Resources of Nova Scotia* (Halifax: A. & W. MacKinlay, 1849), 221-223.

2. Ibid., 248 & 249.

3. Ibid., 223 & 224.

4. Ibid., 249.

5. Ibid., 116.

6. Ibid., 116-117.

7. Ibid., 118.

8. Ibid., 248.

Chapter 8

1. Abraham Gesner, *The Industrial Resources of Nova Scotia* (Halifax: A. & W. MacKinlay, 1849), 103-104.

2. Abraham Gesner, *New Brunswick with Notes for Emigrants* (Saint John: Chubb, 1847), 275-280.

3. *The Industrial Resources of Nova Scotia*, 107-109.

4. Ibid., 111-112.

5. *New Brunswick with Notes for Emigrants*, 280-283.

Chapter 9

1. Abraham Gesner, *New Brunswick with Notes for Emigrants* (Saint John: Chubb, 1847), 334.

2. Abraham Gesner, *The Industrial Resources of Nova Scotia* (Halifax: A. & W. MacKinlay, 1849), 176-179.

3. Ibid., 182.

4. Ibid., 201-202.

5. Ibid., 208-209.

Chapter 10

1. Abraham Gesner, *The Industrial Resources of Nova Scotia* (Halifax: 1849), 229.

2. Ibid., 217. The GMA had a monopoly on mining in Nova Scotia. On February 17, 1845, a petition from Gesner for a lease on coal mines in Cumberland County was read in the Nova Scotia House. The petition was referred to the committee dealing with Her Majesty's government and the GMA. See *Journal and Proceedings of the Nova Scotia House of Assembly* (page 15) for that year. Nothing came of Gesner's request.
Because of this British monopoly, coal miners were brought from the mother country and Nova Scotian-born workers were not usually employed — thus Gesner's observation: "The miners are principably from Scotland." (*The Industrial Resources of Nova Scotia*, 217).

3. Ibid., 267.

4. Abraham Gesner, *Remarks on the Mineralogy of Nova Scotia* (Halifax: Gossip and Coade, 1836), 154.

5. Ibid., 157-159.

6. Ibid., 165.

7. Abraham Gesner, *A Practical Treatise on Coal, Petroleum and other distilled oils* (New York: Baillière Brothers, 1861), 71.

8. Ibid., 89.

9. Ibid., 107.

10. Ibid., 123.

Chapter 11

1. Abraham Gesner, *The Industrial Resources of Nova Scotia* (Halifax: A. & W. MacKinlay, 1849), 218-219.

2. Ibid., 210-211.

3. Ibid., 256-257. For a further understanding of the ins and outs of the closing down of the Annapolis Iron Company, consult Gesner's publication on the subject: *Report on the Annapolis Iron Works* (Halifax: James Bowes & Son, 1853), 8 pages.

4. Ibid., 259.

5. Ibid., 32-33.

6. Ibid., 9.

7. Ibid., 212.

8. Ibid., 16.

9. Ibid., 17.

Chapter 12

1. These excerpts comprise less than half of Gesner's *The Gold Fields of Nova Scotia* (New York: J.P. Prall, 1862). Tangier and "The Ovens" were, however, the districts which, in 1861, occasioned the most interest in Nova Scotia gold.